Using Multimodal Representations to Support Learning in the Science Classroom

Brian Hand • Mark McDermott • Vaughan Prain
Editors

Using Multimodal Representations to Support Learning in the Science Classroom

Editors
Brian Hand
University of Iowa
Iowa City, IA, USA

Mark McDermott
University of Iowa
Iowa City, IA, USA

Vaughan Prain
La Trobe University
Bendigo, Victoria, Australia

ISBN 978-3-319-36232-8   ISBN 978-3-319-16450-2 (eBook)
DOI 10.1007/978-3-319-16450-2

Springer Cham Heidelberg New York Dordrecht London
© Springer International Publishing Switzerland 2016
Softcover re-print of the Hardcover 1st edition 2016

Printed on acid-free paper

Springer International Publishing AG Switzerland is part of Springer Science+Business Media (www.springer.com)

Contents

Chapter 1
Learning Science Through Learning to Use Its Languages

Vaughan Prain and Brian Hand

Focusing on the Medium of Science Learning

Over the last 30 years science education researchers have focused intensively on the role of language, and increasingly the languages of science, to understand how to deepen student engagement and learning in this subject. The medium for learning as well as the content has been foregrounded in this agenda. Partly this focus was prompted by Lemke's (1990) ground-breaking work on the key role of classroom talk in this learning, and the growing recognition that students in learning science had to learn a new literacy that underpinned and characterised scientific realities (Moje 2007). Partly this focus on richer pedagogy also arose from recognition that more effective teaching and learning practices were needed to address continuing widespread student lack of quality performance and sustained interest in this subject (Osborne and Dillon 2008; Weinburgh 1995). This led to an increased focus on researching the role of talk and writing as crucial epistemological tools for science learning (Halliday and Martin 1993; Prain and Hand 1996; Rivard and Straw 2000; Yore et al. 2003).

In this book we present a current international snapshot of classroom-based research on the role of language (and more broadly the languages of science, incorporating multi-modal texts with linguistic, visual and mathematical modes) in learning in this subject. Our intention is to highlight the diverse, generative classroom practices and challenges arising from this broad agenda. We also note the range of research methods used to track and analyse these new practices. Rather than provide a brief summary of the content of contributors' chapters, we overview

V. Prain (✉)
La Trobe University, Bendigo, Australia
e-mail: v.prain@latrobe.edu.au

B. Hand
University of Iowa, Iowa City, IA, USA
e-mail: brian-hand@uiowa.edu

© Springer International Publishing Switzerland 2016   1
B. Hand et al. (eds.), *Using Multimodal Representations to Support Learning in the Science Classroom*, DOI 10.1007/978-3-319-16450-2_1

the emergence of this agenda, its theoretical underpinnings, key recurrent pedagogical and research questions posed by this agenda, accounts of learning processes and learner capabilities, and the research methods used to analyse this learning. In noting these themes we outline the degree of convergence and diversity across contributor perspectives.

Early Focus on Writing

This research over the last three decades has sought to (a) identify what is or might be known and learnt by students through writing in science, and (b) explain how, and under what conditions, this writing promoted learning. In the 1990s two dominant accounts of the role of writing as a learning tool guided this research. The genrist approach (see Halliday and Martin 1993; Veel 1997), drawing on cognitive processing theories (Johnson-Laird 1983) assumed that language organized and represented thought, and that students needed to be inducted into science language practices. In this way, generic knowledge of the form/function of science texts, once internalised by students, "provides the basis for a new disciplined way of seeing and thinking" (Bazerman 2007, p. 8). From this perspective, knowing and reasoning in science depended on students' acquisition of subject-specific writing skills, evident in the writing practices of scientists (Halliday and Martin 1993). By contrast, advocates of a "learning through writing" approach (Sutton 1992; Hand and Prain 1995; Rowell 1997), claimed that to acquire the new literacies of science, students needed to write in diverse ways for different readerships to clarify understandings for themselves and others.

Both perspectives assumed that writing operated as a key epistemological tool for learning, in that drafting and revising processes enabled students to build and review links between classroom activities, conceptual understandings, and their expression. Both perspectives assumed that writing in science was fundamentally about disciplinary reasoning in this subject. The genrist approach emphasized the necessity of fidelity to disciplinary norms of expression for learning to occur, whereas the writing-to-learn approach stressed personal meaning-making through links to natural language and everyday communicative contexts. Both claimed that classroom research using these approaches aligned with the knowledge-production and representational practices of scientists. The genrists focused on representational norms, but this priority left open the question of how student would move from spectators of other people's reasoning to scientific reasoners themselves. The "learning through writing" researchers tended to focus on the need for students to experience first hand the creative reasoning that underpins the production of new knowledge by scientists. However, this left open the question of how and in what ways students would acquire an understanding of how to use experimentation and the symbolic systems of science (vocabulary, graphs, tables, diagrams) to reason persuasively in this subject. Both sides saw that students needed to be active meaning-makers rather than simply meaning-receivers, but differed about what exact learning experiences were to be prioritized to achieve this end.

Genrist Research

The genrist viewpoint assumes that the languages of science are broadly a stable, denotative, representational system that must be learnt in order for students to demonstrate science literacy. According to Martin (2000), Veel (1997), and others, students will learn effectively the rules and meanings of the particular language practices of science through the following teaching strategies: detailed analysis of linguistic features of textual examples; joint construction of genres with their teacher; and through an explicit extensive teacher focus on key textual function/form relationships and their rationale. In other words, researchers within this orientation favour a highly directed, explicit teacher-focused pedagogy that emphasizes the functional aspects of language features of this discourse. Classroom research based on this perspective has largely taken the form of case studies of reputed desirable or exemplary implementation (Martin 2000; Martin and Rothery 1986; Scheppegrell 1998; Unsworth 2001). While this research has established increasingly complex accounts of the tasks learners face in understanding and mastering specific multimodal genres, these studies have not assessed contrasting treatments, and have therefore not established a case for greater learning gains for this approach over others. The evolving nature of functional dimensions of web-based science texts has further complicated genrist attempts to move beyond descriptive accounts of texts to meta-functional principles and workable classroom practices.

Writing to Learn Science Research

Researchers within this perspective, such as Levin and Wagner (2006), Hand and Keys (1999), Rivard and Straw (2000), Rowell (1997), Wallace et al. (2004), and others, assert that students, in striving to clarify networks of concepts in science topics, should be encouraged to write in diverse forms for different purposes. Descriptive studies where diversified science writing tasks have been used have reported positive effects on students' attitudes towards, and engagement with, the subject (Prain and Hand 1996). Comparative studies of contrasting treatments have been conducted by Hand and his colleagues around diversified writing types, including the use of a framework called the Science Writing Heuristic (Keys et al. 1999; Hand and Keys 1999; Hand 2007). This framework of a modified laboratory report leads students through a reiterative process of knowledge construction in science through making and justifying claims, gathering and representing evidence, and reflecting on the progression of ideas. Gunel, Hand, and Prain (2007) noted that using writing-to-learn strategies was advantageous for students compared to those students working with more traditional science writing approaches. In another study Gunel et al. (2004) reported that students' performance in answering higher order cognitive questions was enhanced when students used a modified writing genre, when contrasted with student use of the traditional laboratory report, although the teacher's implementation strategies were viewed as a major factor in this outcome. The researchers claimed that writing serves learning when (a) writing tasks are designed to require students to focus on conceptual understanding, and also require students to elaborate and justify these understandings of the topic; (b) the target readership is meaningful for the students, (c) students are provided with sufficient planning support, and (d) planning activities engage students in purposeful backward and forward search of their emerging texts.

Writing Within Multiple Modes of Representation in Science

Parallel to this research on writing, there has been growing recognition of modal interdependence in science knowledge production and in interpreting and constructing science texts, with an increasing emphasis on visual modes. As noted by Wise (2006, p. 75), much of the history of science has been about "making new things visible, or familiar things visible in new ways", where images "form both what and how we know" (p. 82). In applying this focus on multimodality to the classroom, Lemke (2004, p. 41) pointed out that students needed to "integrate multiple media simultaneously to reinterpret and recontextualize information in one channel in relation to that in the other channels". Students have to translate, integrate and reinterpret meanings across verbal, visual and mathematical expressions, as well as connect these modes to earlier experiences of science activity. This is evident when students interpret the individual and relational meanings between a diagram, an accompanying text, and its referents in the world. Equally, students participate in similar processes when they construct their own text to clarify or elaborate on the meaning of an accompanying graph, photograph or diagram. For Lemke (2004, p. 2), writing's forte is its capacity to enable "reasoning about relations among categories" because it operates primarily by categorical contrasts and exclusions. Quantitative meanings such as rates and angles of change, and alterations to shape and motion, are more suited to visual and mathematical representation. In this way, Lemke argued that science is necessarily about reasoning across interdependent modes of measurement and explanation. He further argued that the use of natural language, and by implication writing, enabled links to be made between qualitative observation and linguistic reasoning about verbal categories, concepts, and their justification.

However, in the messy world of the classroom there are often significant challenges students face to achieve shared well-founded understandings of target concepts and scientific reasoning processes and resources. In commenting on the epistemological role of language, and by implication writing in learning, Anderberg,

Svensson, Anderberg et al. (2008) argued for the dynamic and ambiguous character of the relations between students' meanings, conceptions and expressions. They noted that reproducing disciplinary language does not ensure disciplinary understanding, and that students' intended meaning for an expression is often arbitrary, associative and contextual rather than convention-dependent. Anderberg et al. (2008) asserted that for language to serve learning students must reflect explicitly on the adequacy of the links they are making between intended meanings, conceptions, and different or diverse expressions. These researchers further noted that this use of language as a knowledge-constituting activity is a developmental recursive process. Students need to reflect on the ways they change or develop intended meanings, and to recognize the same meaning across different contexts, different conceptions, and different expressions and modes. These researchers further asserted that students are likely to proceed through a sequence of understandings that starts with isolated local lexical meanings, and superficial relationships between meaning and expression, and develops into more holistic, integrated linkages between concepts, their expression and their referents. By implication, the capacity for student writing to function as an epistemological tool depends on the robustness and coherence of these links. More broadly, the capacity for students to reason multi-modally, to make links between concepts, practical experiences, and their re-representation in writing and other modes, also depends on these linkages. However, what is relatively easy to theorize presents significant challenges in classroom practice.

Chapter contributors in this book draw on this broad mix of theoretical perspectives and diverse foundational starting points to research classroom practices that seek to promote student multimodal reasoning. Some contributors build on insights from genrist perspectives on writing requirements (see Tang and Ho, Tolppanen et al.), while others draw more on sociocultural accounts of learning as context-dependent induction into particular roles and purposes as science learners (see Linebarger and Norton-Meier, Tytler and Hubber, Carolan). There is convergence around the need for teachers to induct students into the mix of visual, verbal and mathematical resources that are the purpose-built tools for claim-making in science texts. While acknowledging the necessary interdependence of modes for students to make convincing claims in these texts, contributors generally assume that talk and writing are the superordinate modes to generate, judge and organize meaning-making in the classroom. They argue that other modes need to be integrated with (or embedded within) written texts, raising complex practical challenges around effective teaching and learning tasks and sequences. As noted by Gunel and colleagues (Chap. 4), Nam and Cho (Chap. 7), Simon (Chap. 2), Tolppanen and colleagues (Chap. 3), and Villaneuva (Chap. 5), in learning about any science topic, students necessarily need to understand the form/function of different modes, and be able to integrate/embed these modes into a coherent multimodal case to make convincing claims. By implication, the meanings students attribute to their writing depend on the meanings they construct from other modes and their capacity to integrate these meanings with their writing.

Explaining Multimodal Reasoning and Learning in Science Classrooms

Our contributors tend to draw on multi-theoretic lenses to guide and justify their approaches to interpreting student reasoning and learning gains from interpreting and constructing multimodal science texts. These lenses include cognitivist perspectives, where learning is broadly understood as guided problem-solving mental work by individuals and groups to come to know and apply scientific concepts, methods and processes (see McDermott and Hand; Carolan; Nam and Ho; Gunel et al.; Tytler and Hubber). Many contributors also draw on Lemke's (2004) socio-semiotic perspective, where learning is understood as induction into the purposes, affordances and opportunities of the multiple sign systems of science, using every-day language and the domain-specific languages of science to make sense of class-room scientific activities/explanations over time (see Gunel et al.; Tytler and Huber; Carolan; Tang et al.). Tang and colleagues further draw on structural functional linguistics to characterise the complexity of the learning tasks facing students as they re-represent experiences, data, concepts, and processes across modes to integrate/construct scientific accounts of phenomena.

There is broad agreement amongst contributors that students are likely to learn this multimodal disciplinary literacy through purposeful guided immersion in these meaning-making practices. However, this raises further questions about what kind of immersion, what roles and tasks for learners and teachers, are productive and manageable when students' practice inevitably precedes competence in acquiring this new literacy. Our contributors broadly agree that quality learning is enabled when students (a) are motivated to represent and justify causal claims about topics, (b) have multiple opportunities to re-represent, translate, justify and refine understandings through processes of experimentation, collaborative peer learning, consultation, and teacher-guided consensus around representational adequacy, (c) come to understand the form/function of different visual, verbal and mathematical scientific representations, and (d) can integrate these modes to interpret and create convincing textual claims in this subject.

However, this agreement about macro conditions for quality science learning raises further questions about how to optimise student success at the micro level of learning about particular topics. This micro level focus includes teacher choices around key concepts in topics, effective task challenges, sequences and learning experiences, the relationships between classroom activities and their representation, and the choice of representations to be considered and integrated. Other issues include the timing and amount of explicit teaching of form/function relationships in scientific representations. Should a toolkit of representational options be taught first rather than learnt in use (or as needed) on particular topics? Does this depend on the topic and/or the age of the students? Are there general principles or conditions that apply to all student learning about how to engage successfully in multi-modal reasoning in science? By contrast, are different strategies required for different age groups depending on likely student capabilities and background knowledge and representational resources? What learning tasks or task sequences will enable students to understand why and how to integrate modes to practice scientific reasoning? How can learners be encouraged to take up the perspective of an ideas-tester, a creative problem-solver, a multi-modal reasoner about phenomena, rather than a bystander at received, proven "solutions" of others? How should teachers understand and enact their roles to put students on a productive epistemological footing? What are appropriate methods and modes to assess this complex disciplinary literacy learning?

Needless to say, our contributors provide varied and sometimes partial answers to these questions, but they all capture the excitement of grappling with these issues, and are optimistic about the potential value of this agenda. In the rest of this chapter we review some key recurrent research questions that they address.

Recurrent Questions

One major question revolves around how the mode of writing could or should relate to other modes, and identifying conditions that optimize effective multimodal learning. Chapter contributors generally concur with Gunel and colleagues (Chap. 4) that writing is "a powerful tool for the construction of scientific knowledge", provided this writing functions to constitute new student knowledge and is complemented by, or integrated with other modes. As proposed by Tolppanen and colleagues (Chap. 3), writing is the dominant mode for representing understanding, supported by other modes such as tables, graphs and diagrams embedded in the text. These researchers claim that writing is a crucial resource for moving between modes, and clarifying the purposes and claims made by the other modes. Simon (Chap. 2), in reporting on Year 10 Austrian students' attempts to explain the roles of non-verbal modes in published expert accounts of science topics, argues that writing is the key structural building block of this multi-modality. In researching multi-modal construction of science understandings by very young students (grades 1–3), Linebarger and Norton-Meier (Chap. 6) note that verbal language provides the main foundation and organiser for this learning and meaning-making, even when these students often favour visual and gestural means to represent their emerging understandings. They claim that this verbal language is crucial for developing very young students' "representational flexibility". Villanueva (Chap. 5) points out that in the context of second language English learners in South Africa, teachers need to know how to negotiate between these learners' everyday language and the understandings implied in the authorized vocabulary and multimodality of science texts. For Nam and Cho (Chap. 7), Year 8 Korean students' written language is the crucial mode for developing and communicating science concepts, where this mode is necessarily supplemented by other modes.

Other contributors elaborate on the necessity of non-verbal modes in learning and communicating science concepts and claims. Tang and colleagues (Chap. 8) confirm the critical roles of talk and writing activities as organizers of science

learning, but also claim these modes need to be holistically connected to other activities, such as physical manipulation of objects, experimentation, and student drawing of their explanatory claims. They emphasize that that teachers need to make
explicit to students the key role of multimodal connections in developing and justifying any scientific explanation. This increased focus on the affordances of all the
modes used in scientific explanations is also proposed by Carolan (Chap. 11), and
Tytler and Hubber (Chap. 9). For Carolan, (Chap. 11) learners have to view all
modes, including physical enactment as well as symbolic modes, as tools to enable
them to be active constructors of possible scientific explanations. In this way, experience of the physical properties of beads can enable learners in grades 4 and 5 to
construct a verbal and visual understanding of a particle model in the topic of
change of state. For Tytler and Hubber (Chap. 9), students can use artifacts such as
a mini-globe and LED torch to explore explanations of day and night cycle, seasons
and eclipses, and thus come to understand the differences between geocentric and
space-centric perspectives when describing and explaining astronomical phenomena. While guided teacher discussion is crucial to these learning gains, student 2D
visual re-representation of their understanding, appropriately annotated, is also
claimed to be critical to this learning.

Another recurrent question posed by contributors is the issue of which communicative tasks and representational challenges are likely to support students learning
(a) the function of multiple modes in scientific texts, and (b) how to embed/integrate
these modes to make persuasive claims in their own scientific texts. The contributors offer a range of procedures and task to achieve these learning outcomes. Some
authors, such as Simon (Chap. 2), McDermott and Hand (Chap. 10), and Tolppanen
and colleagues (Chap. 3), focus on the need for explicit student analyses of expert
science texts to identify how authors use a range of cohesive ties to signal how
modes are linked to develop a coherent claim or claims. These analyses are then
expected to be the bases of student transfer to their own multimodal text production.
Simon reports on how students in his study were supported to develop criteria for
judging the effective use of modes in published texts, such as the use of illustrations,
diagrams and graphs to attract reader attention, clarify processes, and persuade
readers of the reliability of authorial claims and findings. McDermott and Hand
(Chap. 10) designed a specific lesson to highlight strategies used to embed multiple
modes of representation within writing tasks. Students were expected to generate a
matrix for assessing science texts by how well different modes were integrated and
the degree of cohesion of the text as a whole. Tolppanen and colleagues (Chap. 3)
proposed a similar approach where three classes of Year 8 students in Finland were
expected to analyse published science texts to develop criteria to assess the value of
different modes in these texts, and then apply these criteria to generate an effective
multi-modal text of their own. These criteria included the soundness of claims, logical ordering of modes, explicit cohesive verbal ties between modes, and where textual explanations were in close proximity to relevant non-verbal modes.

Other contributors, such as Carolan (Chap. 11) and Tytler and Hubber (Chap. 9),
tend to focus on the need for students to be given tasks that entail a representational
challenge, where students apply emerging understandings to make a multi-modal

causal claim about some new aspect of the topic, or new context. In this representation construction approach, the students are expected to explain their representational choices, be guided by their teacher to judge the adequacy of their own and peer representations, and reach a class consensus about both (a) what counts as a persuasive claim in this topic, and (b) effective multi-modal representational choices to communicate this claim. This approach implies some flexibility about how teachers focus on form/function of modes, and options to achieve student understanding of how modes can be integrated through cohesive ties. For Carolan (Chap. 11), a key issue is how teachers frame what is required of students as participants in knowledge-making and knowledge-testing activities.

Our contributors also seek to encapsulate what enables students to learn and apply this new literacy to different topics and different levels of schooling. Variously they highlight the value of affordances in the tools and processes (see Tang and colleagues; Tytler and Hubber; Carolan), the necessity of re-representation and translation work across modes (see Tang, McDermott and Hand; Linebarger and Norton-Meier; Gunel and colleagues; Nam and Ho), appropriate fit of modes and task sequences to student capabilities (McDermott and Hand), and explicit instruction on how modal embeddedness and integration are achieved by expert science communicators (see McDermott and Hand; Tolppanen and colleagues; Simon; Villanueva). The chapters collectively point to the complexities entailed in achieving enhanced learning environments for learners of different ages engaging with different topics, and using different resources. Our contributors also draw on various research methods to track and analyse classroom processes and outcomes. These include quasi-experimental, pre-posttest design (see Gunel and colleagues; McDermott and Hand), tight track of learning sequences (see Tang and colleagues; Tolppanen and colleagues), multi-theoretic lens case study (Carolan) case studies (Linebarger and Norton-Meier; Simon; Tytler and Hubber), content analyses of relevant literature (Villanueva), and content analyses of artefacts and contexts (see especially Tang and colleagues; Tytler and Hubber; Carolan).

We are excited by the insights and findings reported by our contributors as they take up this generative lead into quality learning in school science. We recognize that many questions raised by this agenda remain open-ended or in need of more extended examination. However, we present these studies as indicative of the promising scope entailed in this focus on the resources for meaning-making in this subject.

References

Anderberg, E., Svensson, L., Alvegard, C., & Johansson, T. (2008). The epistemological role of language use in learning: A phenomenographic intentional-expressive approach. *Educational Research Review, 3*, 14–29.

Bazerman, C. (2007). Genre and cognitive development: Beyond writing to learn. http://www3.unisul.br/paginas/ensino/pos/linguagem/cd/English/5i.pdf. Accessed 28 Nov 2007.

Gunel, M., Akkus, R., Hohenshell, L., & Hand, B. (2004, April). Improving student performance on higher order cognitive questions through the use of the Science Writing Heuristic. Paper presented at 288 L. M. Hohenshell and B. Hand the annual meeting of the National Association for Research in Science Teaching, Vancouver, B.C., Canada.

Gunel, M., Hand, B., & Prain, V. (2007). Writing for learning in science: A secondary analysis of six studies. *International Journal for Mathematics and Science Education, 5*(4), 615–637.

Halliday, M., & Martin, J. (1993). *Writing science: Literacy and discursive power*. London: Falmer Press.

Hand, B. (Ed.). (2007). *Science inquiry, argument and language: A case for the Science Writing Heuristic*. Rotterdam: Sense Publishers.

Hand, B., & Keys, C. (1999). Inquiry investigation. *The Science Teacher, 66*(4), 27–29.

Hand, B., & Prain, V. (Eds.). (1995). *Teaching and learning in science: The constructivist class-room*. Sydney: Harcourt Brace.

Johnson-Laird, P. (1983). *Mental models: Towards a cognitive science of language, inference and consciousness*. Cambridge, MA: Harvard University Press.

Keys, C., Hand, B., Prain, V., & Collins, S. (1999). Using the science writing heuristic as a tool for learning from laboratory investigations in secondary sceince. *Journal of Research in Science Teaching, 36*(10), 1065–1084.

Lemke, J. L. (1990). *Talking science: Language, learning, and values*. Norwood: Ablex.

Lemke, J. (2004). The literacies of science. In E. W. Saul (Ed.), *Crossing borders in literacy and science instruction: Perspectives in theory and practice* (pp. 33–47). Newark: International Reading Association/National Science Teachers Association.

Levin, T., & Wagner, T. (2006). In their own words: Understanding student conceptions of writing through their spontaneous metaphors in the science classroom. *Instructional Science, 34*, 227–278.

Martin, J. (2000). Design and practice: Enacting functional linguistics. *Annual Review of Applied Linguistics, 20*, 116–126.

Martin, J. R., & Rothery, J. (1986). What a functional approach to the writing task can show teachers about 'good writing'. In B. Couture (Ed.), *Functional approaches to writing: Research perspectives* (pp. 241–265). London: Frances Pinter.

Moje, E. B. (2007). Developing socially just subject-matter instruction: A review of the literature on disciplinary literacy teaching. *Review of Research in Education, 31*, 1–44.

Osborne, J., & Dillon, J. (2008). Science education in Europe: Critical reflections. A report to the Nuffield Foundation.

Prain, V., & Hand, B. (1996). Writing and learning in secondary science: Rethinking practices. *Teaching and Teacher Education, 12*, 609–626.

Rivard, L., & Straw, S. (2000). The effect of talk and writing on learning science. *Science Education, 84*, 566–593.

Rowell, P. A. (1997). Learning in school science: The promises and practices of writing. *Studies in Science Education, 30*, 19–56.

Scheppegrell, M. (1998). Grammar as resource: Writing a description. *Research in the Research in the Teaching of English, 25*, 67–96.

Sutton, C. (1992). *Words, science and learning*. Buckingham: Open University Press.

Unsworth, L. (2001). *Teaching multiliteracies across the curriculum: Changing contexts of text and image in classroom practice*. Buckingham: Open University Press.

Veel, R. (1997). Learning how to mean—Scientifically speaking. In F. Christie (Ed.), *Genre and institutions: The language of work and schooling* (pp. 161–195). London: Cassell Academic.

Wallace, C., Hand, B., & Prain, V. (2004). *Writing and learning in the science classroom*. Dordrecht: Kluwer Academic Publishers.

Weinburgh, M. (1995). Gender differences in student attitudes toward science: A meta-analysis. *Journal of Research in Science Teaching, 32*(4), 387–398.

Wise, N. (2006). Making visible. *Isis, 97*(1), 75–82.

Yore, L. D., Bisanz, G. L., & Hand, B. M. (2003). Examining the literacy component of science literacy: 25 years of language arts and science research. *International Journal of Science Education, 25*(6), 689–725.

Chapter 2
Writing Popular Scientific Articles, Development of Interest in the Natural Sciences, and Non-textual Representations in Student Texts: The "Young Science Journalism" Program in Austria

Uwe K. Simon

Interest in and Attitudes towards the Natural Sciences amongst High School Students

A typical student attitude toward science is that it is either too difficult to understand or boring and unrelated to real-life. Scientists in turn are often perceived as males with little social interaction with others. This prejudice is often encountered when talking to young people, and several studies show that many teenagers have little interest in the natural sciences, or even a negative attitude towards them (e.g. Krogh and Thomsen 2005; Osborne et al. 2009). In the European Union, one consequence of this attitude has been a lack of qualified graduates in science and technology to fill the existing need in the job market (e.g. Bundesagentur für Arbeit 2012; Gago et al. 2004). However, scientific topics seem to be viewed differently by male and female students. Comparing the figures for study choice of German A-level students, Holstermann and Bögeholz (2007) noticed that there was a significant sex-bias for specific subjects. Many more first year students in physics courses at the university level were male. This predominance of male interest was less noticed in chemistry, and biology was the most preferred of the three subject areas for females. A consistent finding was observed when students for each sex were asked to rank their preferences in these three science areas. The greatest number of males preferred physics, followed by chemistry and then biology, while this trend was exactly reversed in females (Holstermann and Bögeholz 2007).

Similar data were found for Austrian students (Statistik Austria 2012) (Table 2.1). However, it is interesting to note that almost twice as many female students compared to male students had enrolled in coursework to become chemistry teachers. One potential explanation for these figures could be the social aspects of

U.K. Simon (✉)
Center for Biology Teacher Education, Karl-Franzens-University, Graz, Austria
e-mail: uwe.simon@uni-graz.at

© Springer International Publishing Switzerland 2016
B. Hand et al. (eds.), *Using Multimodal Representations to Support Learning in the Science Classroom*, DOI 10.1007/978-3-319-16450-2_2

Table 2.1 Number of first-year students in natural science-related subjects at Austrian universities in 2011

	Physics	Physics teacher studies	Technical physics	Chemistry	Chemistry teacher studies	Technical chemistry	Biology	Biology teacher studies	Biomedicine and biotechnology	Molecular biology
Male	211	39	315	213	27	152	468	84	4	129
Female	63	16	71	197	50	101	855	222	15	299
Sum	274	55	386	410	77	253	1323	306	19	428
% of all Austrian first year students	0.65	0.13	0.92	0.98	0.18	0.6	3.16	0.73	0.05	1.02

Source: Statistik Austria (2012)

teaching and the chance to combine job and family. All biology courses are female-dominated, even those focusing on molecular biology and biomedicine/biotechnology. This seems surprising, since these latter subjects are much more technology-oriented and require, at least in the second case, more physics than regular biology and findings from research in this area tend to indicate less preference for technology and physics courses by females. For instance, a report from Holstermann and Bögeholz (2007) found that Swedish, English and German female teenagers were much more interested in human biology and medical topics than males, while the latter preferred to know more about physics, technology and electronics than their female counterparts.

Overall, the number of students who enrolled in natural sciences in all Austrian universities in 2011 was rather low: 2892 out of 41,873, which is 6.9 % of all first-year students. At the same time, Austrian universities counted 3274 (7.8 %) first-degree students inscribed for law, 627 (1.5 %) for sociology, 825 (2.0 %) for business studies, 1355 (3.2 %) for pedagogy, 971 (2.3 %) for psychology, and 1241 (3.0 %) for German or German teacher studies. Thus, with the exception of biology, each of the natural science studies referred to in Table 2.1 was far less attractive than these other fields. However, job prospects, particularly in physics and chemistry are comparatively good, while they are less so in biology.[1] Clearly, the number of students interested in studying natural sciences is not meeting the demand for workers in these same areas, especially in the case of physics and chemistry.

Apart from economic concerns, future generations should be willing to engage in debates about significant societal issues such as nuclear vs. regenerative energy, the chemistry of synthetic products, biodiversity etc. To do so productively, students need to understand the related scientific concepts. In the words of Krapp and Prenzel (2011): "It is not only a question of gaining new blood in the field of science. Science concerns everybody – in both everyday and professional life" (p. 28). Finally, young people should be given the chance to discover science's inherent fascination. Yet all the above studies and figures indicate that there is urgent need to create a much more positive view about natural sciences amongst teenagers.

Science Communication: Writing About Science

Scientists often present their results in a written format. To be scientifically literate, students must understand the characteristics of written scientific discourse. This is even demanded in official school curricula (e.g. Austria: BMUKK 2008; Germany: KMK 2004). However, with the exception of documenting laboratory work writing

---

[1] Based on personal discussions with company and university representatives in all three subjects as well as on a rough screen of internet job offers in Austria: For example, there were 138 fulltime and permanent non-teacher positions open for biologists at http://www.careerjet.at, while there were 582 in physics and 533 in chemistry (in contrast to 314 in law) on October 21st, 2013. This means that with current figures and assumed that all first-year students will obtain their university degree almost every graduate from physics and chemistry, but only every 13th biology student and only every 10th law student will find a job broadly related to his/her studies in Austria.

is almost absent from school biology, chemistry or physics in many countries mainly because science teachers have not been trained for supervising writing in their courses (Leisen 2010). Accordingly, many of them do not feel adequately prepared to instruct and assess students in regard to written communication and many teachers do not consider this a part of their job. Yet beginning with 2015 every Austrian student who wants to pass A-levels will have to write a final thesis – and it might well be that some will want to do so in the natural sciences. This has created a situation in which there is pressure on teachers (and students) to make writing an integral part of final grades, but science teachers (and students) feel mostly unprepared to meet this challenge.

Writing to Interest Students in Natural Sciences: The *Young Science* Magazine

The previous passages clearly demonstrate the need to make studying the natural sciences at the university level more attractive to students, especially to females, and to create more situations to practice writing in science classes. Haste et al. (2008) have shown for 14 and 15 year-olds that females' interest in science was strongly linked to ethical considerations and to whether or not scientific issues were regarded as relevant to their lives. Therefore it seems that "offering a homogenous curriculum to all is a mistake – what interests females being unlikely to interest males and vice versa" (Osborne et al. 2009, p. 6). However, as long as co-education is the prevalent form of schooling in most countries, and as long as teachers are not specifically trained to identify issues related to and plan lessons according to sex-specific needs, teaching at school should relate to concepts that increase the likelihood that students of both sexes will become involved in biology, chemistry and physics to the same degree.

Yet students often encounter in their coursework science-related texts which are rarely interesting to read or even unintelligible to many (Sadoski 2001). This is especially true for chemistry and physics (Leisen 2010; Merzyn 1998a, b, c, 2010, 2013), less so for biology, and may be one possible factor explaining why chemistry and physics suffer a greater lack of interest among adolescents than biology. Typically, with younger students, science is presented less abstractly. Consequently, it is easier for younger children to become engaged with scientific issues. It has often been noted that there is a significant drop of interest in natural sciences during the primary-secondary and the lower-upper secondary transition (see Osborne et al. 2009). The nature of textbooks and classroom language might contribute to this phenomenon (Norris and Phillips 1994).

One potential solution for dealing with both the lack of interest in the natural sciences and the difficulties many students have with finding science learning relevant may be to allow students themselves to select the science concepts to communicate about. Several factors currently constrain this practice, including the following:

- Student choice of topics may result in the selection of concepts instructors feel they have little background knowledge in.
- Curricula constraints lead students to the conclusion that they have little time to engage in additional research and writing.
- For the same reason instructors may be unwilling to utilize instructional time for concepts they consider irrelevant for final examinations.
- As mentioned above, science instructors may feel they have inadequate training associated with improving students' writing abilities.

Although these constraints exist, allowing students the ability to write about what they are truly interested in could lead to an increase in interest in the subject matter of science in general. This increase in interest could also lead to a greater motivation in class. To test these hypotheses regarding student interest and motivation stemming from choice in writing about science concepts the *Young Science* magazine was founded in 2012, which is, to our knowledge, unique in Europe. In this magazine, high school students write about topics from biology, chemistry, and physics. They obtain feedback both in terms of language and scientific content from the editorial board consisting of scientists and teachers from all three subjects. Articles are usually revised several times before they get published. The journal itself is distributed for free due to the cooperation with regional school authorities, funding from university and government agencies, and advertisements. Currently *Young Science* has a circulation of 10,000 copies and reaches about 150 secondary schools in four Austrian federal states. First articles from abroad (e.g. Italy) show that interest in this project is becoming ever greater. Figure 2.1 shows the first three covers of the magazine.

As can be derived from Table 2.2, the number of female teenage authors is higher than that of male authors. Only in the first edition are numbers equal, which stems from the fact that a grade ten biology class consisting almost

Fig. 2.1 Covers from the first three issues of *Young Science*

Table 2.2 Number of female and male student authors (single or in small teams) in each issue of *Young Science*

	Issue 1	Issue 2	Issue 3[a]	Issue 4 (in preparation)[a]
Female authors	7	6	3	4
Male authors	7	2	1	

[a]One additional article was provided by a whole mixed class

exclusively of males were specifically encouraged to write for the journal by its biology teacher. Otherwise, females seem to be more interested in providing texts for the journal. However, their topics are by no means exclusively related to humans or biology in general, but also deal with themes such as photo-acoustic imaging or chemiluminescence. Writing might therefore be one way to attract students and, in particular, females to become interested in the natural sciences or, at least, strengthen an already present but possibly tentative attachment to this domain.

Not every article can be published in such a magazine. Internal school journals or web pages, or, at least, a publication within the class presented to classmates and parents would be cheap and feasible alternatives. As will be shown later, the option to tell others about a self-chosen scientific topic is a great factor to motivate students to write such a text.

Apart from interest development, such article writing and publishing may also be seen in a larger pedagogical context to help students develop the writing competence they need to deal with scientific issues in the written format and to learn how to plan and structure such texts. This, of course, includes multimodal representations, since scientific articles use illustrations and tables to both attract the reader's interest and visualize and explain results and ideas. Students need to understand that pure text is unattractive to read for many, and that carefully selected or self-made visual aids may significantly contribute to a reader's understanding of a specific topic. In this respect our authors are encouraged to contribute pictures, graphs or other kinds of additional representations themselves, either by their own material or by asking for re-use of already published illustrations. This is different to the practice of many newspapers and magazines, but prepares for the requirements of scientific publication. By diligently choosing or creating suitable non-textual representations, students will be engaged with their topic from a different perspective: how to best present their main message and/or results in one or few key illustrations or tables. Consequently, this kind of article writing and publication trains students to compose scientific texts and helps them to understand how written scientific communication works by combining both text and non-textual modes. This will be even more efficient if the products of such efforts are published and discussed in class, giving the authors the feedback they need to improve their writing and choice of non-textual representations.

Research Projects to Study Interest Development in and Attitude Change towards the Natural Sciences after Writing Popular Scientific Articles

Creation of Interest

In school, students usually do what they are told to do. Very rarely are they given options to choose between. This might severely limit interest development. According to Krapp's "person-object-theory of interest" (POI) (Krapp 2005), the arousal of interest requires an "ongoing process of person-object interactions." However, there are certain preconditions that are typically necessary to develop interest in something. First, the task and/or goal must be regarded as meaningful and important and must cause some kind of positive emotions (Krapp 2005). Only then will interest become long-lasting as opposed to superficially and momentarily maintained, and only then can a "domain-specific situational interest and later a relatively stable individual interest of high personal relevance" be created (p. 383). Consequently, for students to become interested in natural sciences through writing about them, not only the topic but the writing itself needs to be considered as meaningful/important and linked to some kind of positive emotion. In the context presented here, the assumption was that students saw the writing they were asked to create as both good practice for the requirements of the final thesis and their later career (meaningfulness of task), and that the text they composed would create some sort of pride (positive emotion).

This alone, however, might not be sufficient. Decy and Ryan (1993, 2008) postulated that learning works best when three psychological basic needs are fulfilled: the feeling of competence, autonomy, and relatedness. Accordingly, students should realize that they can master the task, even if it is very new and possibly difficult for them – such as article writing in science classes. Secondly, students need to experience some kind of autonomy without having the teacher telling them exactly what to do and how. In terms of article writing they should thus be given a certain degree of freedom regarding topic choice and approach. Finally, students need to recognize that the task and its content are not only meaningful to the teacher and school curriculum, but also to others, ideally their peer group. For the program presented here this means that students realize that natural sciences have so many interesting aspects to offer that others (classmates, *Young Science* teenage authors from various schools) find them exciting enough to write and/or read about.

Measuring Interest Development

Interest development due to writing was measured through pre- and post-tests. In order to create the questionnaire to be utilized to measure this concept, it was necessary to determine how student "interest" and "attitude" could be dealt with as theoretical constructs. Consultation with the literature proved inconclusive. While both constructs are often treated as separate (e.g. Ellis and Gerberich 1947; Gardner 1996, 1998; Krapp and Prenzel 2011), other authors are less strict in their distinction (e.g. Christidou 2011; Osborne et al. 2009; Schreiner 2006; Vogt 1998; Vogt et al. 1999). Consequently, Schreiner (2006) concludes that "the boundary between them is still blurry" (p. 29). The situation is even more complicated by the fact that both "interest" and "attitude" themselves have been understood and defined differently by different researchers (cf. Krapp and Prenzel 2011; Schreiner 2006). The assumption for the exploratory project was that students would not see interest and attitude as something different, but rather be interested in a task or a topic which they would have a positive attitude towards. Consequently, both constructs were regarded as synonymous.

Concept

Following these considerations a workshop-based research program entitled "Young Science Journalism" was developed to test the idea whether writing about science and the use of an audience outside of classroom peers might increase students' interest in the natural sciences. In this program high school students write and revise popular scientific articles in class with both their science and their language (German) teachers participating. For their writing they also have to heed certain standards of communication about scientific concepts such as in-text citations. The effect of this intervention in terms of interest and attitude development is tested by means of questionnaires and interviews. In parallel, the development of writing competency of students is monitored by comparison of the different revisions they have to hand in after receiving feedback for their respective drafts. In the following sections, an overview of the pilot project is presented. In addition, data collected in regard to interest development and the use of different representations are presented and analyzed. Finally, the follow-up project which is currently being implemented will be introduced.

Pilot Project 2011/12

The program started with an exploratory project in the school year 2011/12. One grade 10 class with 7 females and 13 males aged 15 to 16 and their biology and German teachers (both female) participated. Students were introduced to the project in October 2011 and taught where to search for information and how to cite sources correctly. The criteria that would be utilized by students to rate the quality of popular scientific texts in newspapers, magazines and subsequently their own articles were discussed in class by the students, the project team, and an invited journalist during the first workshop. This list was slightly revised by the project team after the workshop and finally included "informative and attractive title", "examples from scientific work", "use and explanation of scientific terms" among other criteria. Students were given a recent newspaper article about feather-wearing dinosaurs and had to find and discuss a title. Then, they analyzed this and another (more complex) text according to the criteria collected in the workshop. As an opportunity to focus on characteristics most critical in creating effective scientific communication they were asked to find a topic of interest and to start researching this topic. A first draft had to be handed in six weeks later. This draft was read, analyzed and corrected by the project leader and the teachers. Topics had to relate to biology, since both the science teacher and the project leader were biologists. However, within the domain of biology, students had the freedom to choose their particular area of interest.

During a second workshop in January 2012 students were given anonymous copies of their classmates' drafts and asked to provide feedback about another student's article by using the criteria list, and then changing roles. Afterwards, students received individual feedback from the project leader and the teachers. The students were then asked to revise their texts. These revised texts were again read and corrected during another revision phase after a third workshop in March 2012 in which they received feedback from the project team and from a journalist. A final meeting in class ended the project in May 2012. During this lesson texts were read out on a voluntary basis and the project was discussed in retrospective. In the second workshop students were also encouraged to include non-textual representations in their articles, if they had not already done so.

Before, and after, the project students filled in questionnaires with items concerning their interest in natural sciences, reading and writing. Three females and three males from different achievement and motivation levels were additionally interviewed between workshop one and two and at the end of the project to gain further insight into how students approached this task, which problems they encountered and how they tried to solve them, and how they liked the concept of the project. Achievement level was based on biology grades from the past and current school year, motivation was judged by their biology teacher according to general engagement in class. Interviews were transcribed and analyzed using qualitative content analysis according to Lamnek (2010).

Results and Discussion

With very few exceptions students were quite committed to the project. Data from the interviews and teacher observation showed that females invested on average much more time and efforts than males. The second questionnaire allowed students to note what they liked and what they disliked about the project on a voluntary basis. Additionally, they could offer suggestions for future projects. Overall the project was rated positively by almost all students but with emphasis on different aspects. For example, four males referred to the preparatory effect for thesis writing required at the end of their school career ("practicing scientific writing"). No girl mentioned this aspect. Conversely, three females but no males highly valued the intensive text work and the social interaction within the class during workshops. One Kurdish girl explicitly wrote that she would want more of this kind of activity to happen at school ("write more such texts"), since she felt it had helped her to tackle her writing difficulties. This clearly shows that the writing of articles was regarded as meaningful by members from both sexes (males: goal-oriented in terms of preparation for thesis, females: the learning effect itself). Furthermore, several participants had chosen topics they somehow felt personally interested in: One boy, who had a pet, wrote about the relationship between animals and humans. Another, who was more economy-oriented, presented a text about bituminous rocks and fracking. Two dog-loving females wrote about aggression of dogs, and one girl invested much time and effort in an article about Anorexia nervosa. One of her friends suffered from this disease, and she refined her text several times until it was ready for publication in the magazine. It may thus be assumed that this task was attached with positive emotions in that the topics many students chose to write about were important to them.

With respect to writing competence the German teacher noted an improvement for slightly more than half of the class when comparing essay writing (e.g. presentation of arguments) before and after the project (text composition was not an issue of regular German classes during that period). Ten students also wrote in the post-test that they felt their writing had become better through project participation.

Interest in the natural sciences was measured with several direct and indirect items. For example, students were asked how much they liked reading scientific textbooks, non-fiction textbooks, and popular scientific articles. It was encouraging to see that the interest in reading scientific textbooks had improved for two females and four males, even though it deteriorated with one girl and two males. A similar trend was found for non-fiction literature. The most dramatic effect was reached for popular scientific articles (articles from newspapers, popular magazines and magazines equivalent to "Scientific American" or "National Geographic"). Here, three females and four males were more positive about reading such texts at the end of the project than at the beginning, one boy even much more, while the development was negative for only one from each sex. This kind of text format was apparently new to most of the students and is presumably used seldom in school science. However, these data indicate that it seems a worthwhile tool to raise students' interest in the natural sciences.

One striking result of this study is shown in Fig. 2.2. On average, males had regarded natural sciences as exciting at the start of the project (black line in right part of graphic) and experienced, on average, little or a slightly negative change through project participation (dotted line) eight months later. On the other hand, females had started off with ratings well below those of males (black line in left part of graphic). But while no girl was less interested in natural sciences at the project end, four of them improved in their interest development (upward arrows). Thus, the average value for females reached the level of that for males in the post-test (dotted line). This indicates that the writing approach chosen here was particularly successful with female students, mirroring their greater time investment and efforts (see above). The fact that ten males showed no (no arrows) or even a negative development (downward arrows) concerning interest in natural sciences also deserves special attention: For example, student 10 was a very good and ambitious student. However,

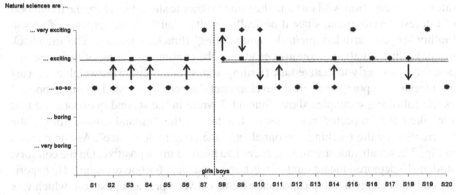

Fig. 2.2 Interest in natural sciences in October 2011 (*diamonds*) and May 2012 (*squares*). The first seven students are females, the last 13 males. In this item, students could finish the prompt "Natural Sciences are ..." with "very boring", "boring", "so-so", "exciting", or "very exciting". *Black lines* denote average values obtained from the first questionnaire, *dotted lines* those from the second. *Arrows* show individual changes

his text failed to meet the high approval he apparently felt he deserved even though he had invested relatively much effort in his article. His remark on the second questionnaire shows his dislike about the several diverse comments he obtained for his text. He completed the prompt *I did not like about the project* with the statement "that some text parts were corrected by one teacher and the same parts by another during the following workshop". A similar statement came from student 12 who had shown almost no difference in any item between pre- and post-test. He wrote: "This is not mine anymore", probably meaning that too many suggestions for improvement had taken the ownership of the text away from him. This perception of loss of ownership was particularly disheartening in light of the relative strength of his text. These two cases seem to be in accordance with the self-determination theory (SDT) of Decy and Ryan (1993, 2008) in that the confrontation with too many external ideas for improvement may destroy students' feeling of competence and autonomy. For these two students the project had thus failed to raise interest in the natural sciences and even diminished interest for one.

In school, texts are usually corrected once to let students know their mistakes (and sometimes even what they did well). Very rarely are students encouraged to revise their texts. It is an essential part in our program to make students work on their texts repeatedly based upon feedback they receive either from teachers and project team members or from peers. In this way, they are encouraged to deal with the same task several times, which should ideally increase learning effects concerning writing competence and also focus students' attention more on the topic they write about than with one text encounter only. This may also be seen as a model for scientific communication, since presentation of results in the written format is usually revised several times before publication, with texts being read by co-workers first and (mostly) anonymous reviewers second. According to Krapp (2005) this intensive occupation with one and the same task/topic should increase the likelihood of interest development, since it basically results in an "ongoing process of person [author]-object [article/topic/task] interactions" (brackets added). The drawback might be that negative emotions because of the continuous facing of the same and possibly not overly attractive task (writing and revising) could outweigh or at least counterbalance positive interest development. Indeed, in our study, both happened as the following examples show. Student 2 wrote in the second questionnaire that she liked the "repeated revisions of the texts", the "mutual commenting", the "correction by the teaching personnel" and the "free topic choice". As can be seen in Fig. 2.2, her attitude towards sciences had become more positive. On the contrary, student 13 deplored the amount of work required "just for one revision." He experienced little change in all questionnaire items between pre- and post-test, which was also true for his interest in natural sciences (Fig. 2.2). Of course, the small number of pupils in this qualitative study prohibits generalizations. Other factors such as identification with topics or positive attitude towards text work in general will additionally come into play here. Nevertheless it is interesting to see how well theory fits for many individual cases.

The interviews were devised in a way to further understand how students tackled this task and what they thought about the project. For example, students were asked what made them choose their topic. In order to facilitate topic choice, the biology teacher had made available a wide range of magazine copies, which were the German equivalents to *Scientific American* and *National Geographic*. Yet only two of the six interviewed students relied on thematic stimuli offered by their teacher. The chance to do something on their own and to choose a topic in some respect relevant to them outweighed the ease of getting started with information given to them. During the second interview session no student mentioned dissatisfaction with his/her choice. However, almost all of the interviewed students needed a long time to decide upon their topics; two said that this process was *"very difficult"*. One (student 3) explained: *"I am not so much interested in biology, rather in medicine. … (Thus I) have tried to choose something in between."* Interestingly, this indicates that she seemed to regard medicine as something separate from biology, even though medical topics belong to the biology curriculum at school. Student 1 described the topic finding process in detail:

> S1: *"It was very difficult. I needed much time to find a topic screening magazines, the internet … I had searched for quite a long time, but finally I thought of stress hormones. Then I spoke with my teacher and she said that this would be a very demanding topic and gave me a book from year 12. I read in the book, and also in the internet, and I did not succeed in getting what I wanted. I had great problems to find relevant information and to understand scientific terms."*

When asked about reasons for topic choice, students mentioned relevance for themselves or suggestions from family members.

> Interviewer: *"The topic stress hormones seems a rather difficult one. Why did you choose it?"*
> S1: *"We shortly mentioned it in biology, and I am interested in stress, its origins…"*
> Interviewer: *"Is this a topic which is personally relevant to you?"*
> S1: *"Yes."*

S3: *"I read some articles in the magazines our teacher gave us. And then I found one about dolphins and their brain, and one about aggressive dogs. Since I am very animal-loving, I found these topics attractive. Then I thought: One of my relatives has a dog which has such problems with other dogs. It is always immediately aggressive towards other dogs".*

S9: *"First I did not exactly know what to do. I did not find anything interesting in the material our teacher presented. Then I asked my aunt. She knows a lot about biology. She gave me a book about animal behavior. How the mantis is eating her partner and stuff. … There was also something about play-dead behavior, and this I found somehow interesting."*

Interviewer: *"What did you find interesting about this topic?"*

S9: *"Well, I don't know. I just had a look and thought: This seems quite interesting. That animals just fall to the ground, that is somehow … I think this is clever."*

Interviewer: *"That they fall to the ground?"*

S9: *"Yes, they pretend to be dead, when they are attacked. And then, then the other animal goes away, the predator. And this I just found interesting."*

Once a topic is found the writer has to find the motivation to do all the research and (re-) writing. One rationale behind the *Young Science* magazine is that good texts are rewarded with a publication available to many thousand students of similar age. Yet even if publication may occur only at a very local scale, (e.g. the school web page) it will require some investment to make this happen. Consequently, one aim of the interviews was to examine, whether the chance to publish their work might have a positive influence on students' motivation. For example, it is possible that students found it stimulating to compose a text not just for their teacher's eyes, but for students their same age. Alternatively or additionally, some authors may develop pride in seeing their own text product in a magazine widely distributed. Two exemplary answers relative to this issue from the first interview session are provided:

S3: *"Somehow you do a bit more and you think, of course, somebody might read it, and perhaps somebody you know and who gets an impression from me."*

S9: *"(the option to publish the text) motivated me to revise it (…) and think about how to improve it. It would be cool, if we were in there, in the magazine."*

Five out of the six interviewed students were very positive about this option. Several answers showed that the idea to indirectly get into contact with others (readers: mostly teenagers from other schools) was an important factor for taking up with revision work for both sexes. This again is a nice proof of the applicability of the SDT (Deci and Ryan 1993, 2008), since it stresses the need for social relatedness of a task and/or its product to be regarded as interesting.

Illustrations in Student Texts

Representations other than text are important parts in scientific papers. Indeed, one may assume that after reading the title and abstract the scientific reader most likely glances at the illustrations in the results section before deciding whether or not to study the main text in detail. Facing the ever increasing number of publications this is probably the only strategy to survive the enormous input of information.

The following two recommendations show how students and young researchers are advised to get a first impression of a scientific paper:

> Rely on the figures to impart the bulk of the information. So to get the most out of the results section, make sure to spend ample time thoroughly looking at all the graphs, pictures, and tables, and reading their accompanying legends! (http://www.sciencebuddies.org/science-fair-projects/top_science-fair_how_to_read_a_scientific_paper.shtml)
>
> Indicators of the author's main points at document level: title, abstract, keywords, visuals (especially figure and table titles), first sentence or the last 1–2 sentences of the introduction. (Purugganan and Hewitt 2004)

It is likely that similar strategies are utilized by students when they encounter information in textbooks or other sources. A limited number of studies have examined the way textbooks, task sheets or examination tests use multiple representations to present scientific information (Donnelly 2010; Nitz et al. 2014; Roth et al. 1999). Other groups have analyzed or discussed how such multiple representations might affect student learning (Ainsworth et al. 2002; Ainsworth 2006; Stern et al. 2003; Tytler et al. 2013; Waldrip et al. 2006; Yore and Hand 2010). But little attention has been paid to exploring why and how students select the modes that they do and what strategies they use to integrate modes (Ainsworth et al. 2011). In the following sections, it will be shown how the high school students participating in the *Young Science* pilot project integrated pictures, tables and graphic elements into their texts and whether these other forms of representation were chosen (or made) to complement the information in the text, presented redundant information, or had little connection to the overall topic.

Students were encouraged, but did not have to include other modes outside of text in their articles. Four different additional types of representation were observed in student writings: pictures, graphic illustrations, glossary boxes (a separately marked text field containing explanations of specific terms), and tables. Twelve of the 20 students (two worked in a team) used pictures to illustrate their articles, with eleven having one photograph, and one having two photographs in their texts (the second, however, without any additional informative value). Two students included a table, one student included two tables plus a glossary box. One student embedded a glossary box to explain a specific term but had no other visual representation. Two students added a graphic element, one of them with an additional table explaining specific terms, and two students did not include any other representation apart from text at all. The following sections analyze student use of non-textual representations in more detail.

Tables

All the tables and glossary boxes were truly complementary to the main text body in that they provided information not found elsewhere. In this respect they were similar to tables in real scientific articles, since only a fraction of the main results of all the data displayed there are normally referred to in the text. One student listed and explained what should be part of a healthy diet. Another student writing about depression created a table to give information about possible causes for the development of this disease. In all cases, tables were added at the position where their information was dealt with in the main text. In other words, tables were used to provide additional information and for illustrative means. Since they were all text-based, they can be considered as part of the main text presented in different format. Not a single table presented data as numbers.

Pictures

As can be expected, most pictures were taken from the internet. It would have been very difficult for students to produce relevant material themselves for topics such as bituminous rocks or Megalodon (a pre-historic shark). However, three students did use their own photographs. Their topics were: aggressiveness of dogs, Anorexia nervosa (an eating disorder), and relationships between animals and humans. All of the overall 12 photographs were purely illustrative without any information not already found in the text. Seven of those correlated well with the main message of the text. Five, however, did not seem to fit well. For example, one boy writing about how bees use different kinds of dances for communication had chosen a picture of a bee sucking nectar from a flower. A girl writing about aggressive dogs showed a German shepherd pacing peacefully on a lawn. It appeared that for many students there was not a spontaneous recognition of the need to connect the different modes of information to the text. This seems a field where further work is needed, since embeddedness of different representations is vital for learning, but "least likely if the visual adjunct to printed words is decorative" only (Yore and Hand 2010, p. 96). Within this exploratory project, there was no instruction or discussion about how to improve multimodal use of elements other than text, because we wanted to see if and how students would integrate such elements without external influence.

Graphic Illustrations

The two graphics occurring in the articles also did not provide any additional information but were solely used to visualize processes referred to and explained in the main text. One girl made a drawing illustrating microinjection, e.g. the moment when a glass needle injects a sperm cell into an egg cell (topic: *in vitro* fertilization and surrogacy). The other student chose a graphic representation which she took from Wikipedia. Writing about the inheritable disease Chorea Huntington she wanted to illustrate the term 'autosomal dominant', which was certainly the most complex concept in her article (Fig. 2.3).

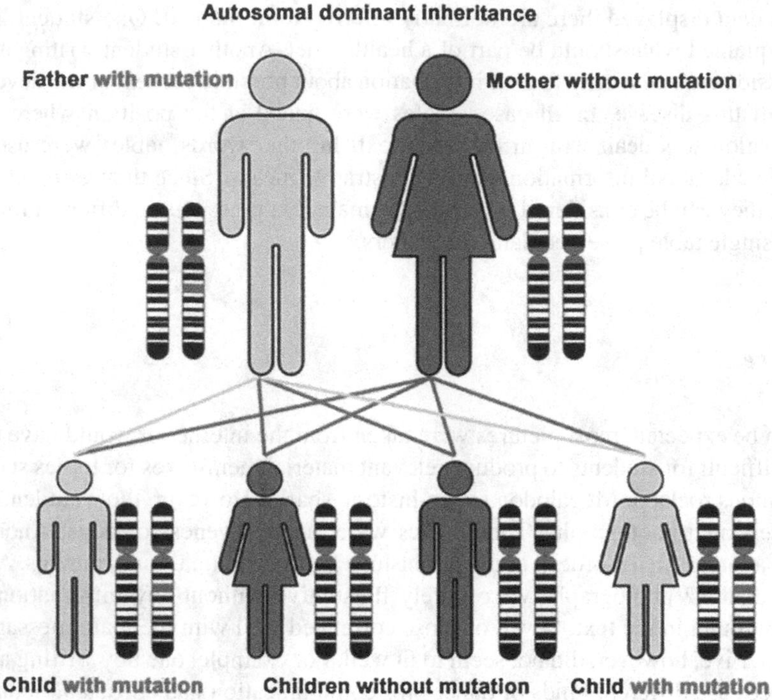

Autosomal dominant inheritance

Fig. 2.3 Autosomal dominant inheritance of a mutation (Source: Wikipedia, modified)

Superscripts, Subtexts and References

Only six students added specific texts to tables, figures, or illustrations, and none but one referred to either of those in the main article. This is clearly a point for further improvement as contrary to purely journalistic texts it is an absolute requirement for scientific texts to provide links between complementary representations and to offer the reader a short summary of the kind of information presented in figures and tables. In journalistic pieces pictures and graphics are often decorative. They lend visual support to what is being written in the text and have to attract the reader's interest. Subtexts may but do not have to relate to key points in the illustration. Conversely, scientific illustrations should not only attract attention, they should also contribute information otherwise hardly describable or even understandable. Thus, in contrast to many illustrations in mass media, scientific illustrations and tables are very often indispensable aspects of the communication of scientific understanding. However, because scientific texts are so dense with information, their readers need guidance to link a particular non-textual representation to the relevant text passage (as is done in this paper by references to particular figures and tables). These requirements also hold true for written assignments at the university level and thus clearly need further attention at all levels of schooling.

In summary, the majority of students used additional representations to illustrate or complement main aspects of their articles. However, only few were creative in the sense that they produced such visual aids themselves, while most used photographs or tables from their source literature or the internet. Due to the vast variety of topics we did not venture to study learning progress in this exploratory project. However, future work with more restrictive topic choice is needed to study whether encouraging students to use their own material contributes to an enhanced understanding of and identification with the topic they write about. Furthermore, the construction (and thus understanding) of diagrams in particular might be an important tool for the diagnosis of students' misconceptions or failures of understanding (e.g. Lachmayer 2008; Nitz 2013).

Follow-up Project 2013/14

In a follow-up project we are currently testing at a larger scale whether writing an article about a self-chosen topic and repeatedly revising texts according to peer and expert comments influences development of interest in the natural sciences, and how students assess each other's use of illustrations. Eight grade 10 classes from five schools in one of the largest cities of Austria, with a total of 86 male and 80 female pupils, are participating in the project, with one additional control class in each school and several other control classes from further schools which do not receive the intervention treatment. Similar to the procedure explained above, students participate in workshops, give feedback to each other and receive feedback from project team members and their teachers. However, the students also receive additional information about how such texts should look like using both professional and student texts as examples. This aspect of the program was added because experience from the pilot project showed that many students had wished for more intensive instruction. While rewriting their first draft exclusively according to peer feedback, the second revision is based upon comments from the project team and teachers.

All classes fill in questionnaires at the beginning and at the end of the project concerning interest in natural sciences with refined items from the first project to which questions from PISA (2006) have been added (Haider and Schreiner 2006). The current project is implemented in physics, because research indicates this subject suffers most from lack of interest amongst students and, in particular, females in high schools (cf. Merzyn 2013). The overall theme is "energy", since this is a central topic in Austrian grade 10 physics. However, students are given the option to choose between a biological, a chemical, and a physical approach. Thus they are given a high degree of freedom concerning how they want to deal with "energy" in their articles. We are especially interested in determining how many students actually choose a more physics related approach (as opposed to topics stemming from biology or chemistry), if this depends on sex, and whether or not the general interest in physics is elevated even with those students composing more biology- or chemistry-centered articles.

In parallel, using multiple-choice pre- and post-tests with items covering a broad range from energy-related topics dealt with at school we test the understanding of and knowledge about these concepts to see, whether topic choice or article writing about energy in general has an influence on knowledge acquisition. As shown by Hand et al. (2007) for molecular biology, article writing may have a significant impact on learning.

Finally, both the physics and the German teacher of each class receive training in text analysis using the same criteria as the students in their peer reviewing process. This is done to help teachers monitor students' development of writing competence across the different versions they have to hand in and thereby use their new expertise as raters for the project. Secondly, teachers are made aware of central aspects they will be asked to look at when supervising theses in the coming years.

Based on the data and on student and teacher feedback from the first exploratory study, several adjustments were made in the workshop design:

– The number of lessons needed was cut from 16 to nine distributed among three workshops and one lesson for final discussion to diminish the time taken away from other school lessons.
– In the second workshop students did not get feedback concerning their texts from the project team and teachers to avoid the danger that students feel overruled by too many suggestions. Here, students reviewed anonymous copies of their classmates' articles using a comprehensive formal review sheet (Fig. 2.4) and discussed their impressions with each other being subsequently reviewer and author after disclosure of the authors' identities.
– In the third workshop anonymous texts from other schools were presented and analyzed to show students examples of good articles and texts which still need improvement. Here, typical features and mistakes were discussed to provide students with suggestions for their own texts. Furthermore, students were individually informed about content errors in their texts by project team members and teachers, since it cannot be expected that their classmates will notice such, and received feedback on text structure and presentation.
– During the third workshop, several students were interviewed concerning their experience with the peer review process. This included talking about the use of non-textual representations in science and student articles.
– There will be one interview session with selected students at the end of the project, but with more detailed questions to understand task approach (e.g. How did you get your topic? Was your article a point of discussion at home? How did you handle terms you did not know in your source literature?) and interest/attitude development (e.g. Did your interest in biology, chemistry, or physics change

Review Sheet

➜ Grade the article by awarding points for each criterion (see below).
➜ Give reasons/examples for each rating. If you have suggestions for improvement, please write them down as well.
➜ Underline those five criteria which are most relevant to you.
➜ In case you want to include further ideas, please do so in the last line.

Title of article:

Name of reviewer:

Criterion	Score	Reasons/example(s)
STRUCTURE of article/FORMAL criteria		
1) Title: present; arouses interest and steers reader to content?		
2) Introduction: present; arouses interest and steers reader to content?		
3) End: Conclusion/take home message present and comprehensible?		
4) Structure:		
- clear and comprehensible?		
- one main idea per paragraph?		
5) Cohesion: linking words/expressions in/between paragraphs present?		
6) Coherence: connection of content between paragraphs; always reference to overall topic?		
7) Sentence structure: clear?		
8) Literature: correctly provided?		
9) Sources: re-phrased or quoted (good) or just included without special references (bad)? Reference directly after relevant information in text?		
10) Special terms/scientific language: present?		
11) Special terms: explained?		
12) Glossary: present?		
13) Illustrations: e.g. graphics, pictures? Explained? Sources provided? Reference to illustrations in text? Relation to topic?		
14) Style: objective/subjective, emotional, exciting...		
TOPIC		
15) Topic/main points? What exactly does the author want the reader to know? Is this in concordance with the title?		
16) Topic:		
- comprehensibly presented?		
- sufficient information?		
17) References to scientific studies or about scientists working at this topic: present?		
18) Examples: present and comprehensible?		
19) Significance for daily life: present and comprehensible?		
20) Personal comments		

➜ Grading according to the following scores (for questions 14), 15) and 20): please enter own thoughts):
    • present and very well done
    • present and done quite well, but still some room for improvement
    • to some extent present but much improvement required
    • not present

Fig. 2.4 Review sheet students used for analyzing each other's articles

through project participation? Did your interest concerning "energy" change during the project? Did your attitude towards writing in the natural science change through project participation?).

First Results

Initial data from this study indicate that sex-related distribution of interest-/attitude is similar to results from the pilot study: Females were significantly more positive about general reading and writing, while there was no difference with reading scientific books or articles. In this second project we did not ask for interest in natural sciences in general but differentiated between subjects both with own items and questions from PISA. Here we found that males were generally much more interested in physics than females (Fig. 2.5), while the latter ranked topics of human biology much higher than males which is in accordance with previous research (Holstermann and Bögeholz 2007). There was no significant difference between sexes regarding other subjects such as chemistry, geology or astronomy, and

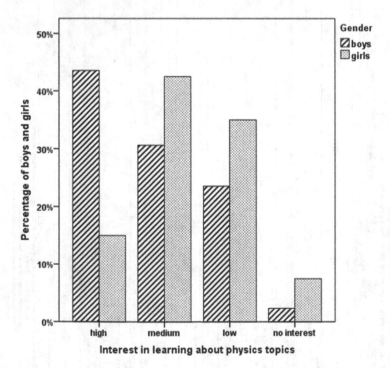

Fig. 2.5 Interest in learning about physics (answers from 85 male and 80 female grade ten students). Males are significantly more interested in physics than females (Mann–Whitney-test, p < 0.001)

concerning NaSc-related PISA items such as "wanting to know how scientists design experiments" or "wanting to study a NaSc subject" etc. Since the project is still running we cannot assess whether or not the writing-based concept presented here will be successful for physics classes, too, and whether it will influence student understanding. Also, we do not have all data from the final interview session yet. However, a clear majority of students participating in the project were highly engaged during the workshops, and many texts are promising. With respect to illustrations similar observations could be made as from the pilot study. This time, however, we asked students and their student reviewers to assess quality and suitability of representations others than text in the articles they read (these data are fully present and have already been analyzed).

Interview Data Relating to Student Use and View of Illustrations in Scientific Texts and Their Own Articles

As in the exploratory project, students were encouraged to add illustrations or other non-textual elements to their text. In addition, they were shown examples of how to best do this. To assess what students thought about illustrations in science texts and how well they believed illustrations chosen by their classmates were integrated with their articles, 56 students (25 male, 31 female) were interviewed a few weeks after they had reviewed one text of another student from their class and had received peer review themselves. We chose students who had either invested much in the revision of their draft after the peer review workshop, or had done little to no work. Interviews were analysed using the freely available software QCAmap (www.qcamap.org).

As can be derived from Table 2.3, almost two-thirds of student judgements about their classmates' illustrations or their number or placement were negative, only 15 statements were explicitly positive. Amongst the first, many critical remarks focused on the lack or small number of illustrations. About one fourth of all interviewed students had the impression that the content of the illustrations did not sufficiently relate to what was written in the text, or that it was purely decorative. Here are some examples:

> Well, yes, it did have some relation to the topic, but there was not that much...with respect to understanding the topic better, ... well, it did not help to make the text easier to understand, just somehow to glance at.
> When it's all about electricity as an alternative kind of fuel it does not help much, if you see a gasoline pump.
> Yes, he put in a picture of, of a perpetuum mobile, just because it was a perpetuum mobile. But he did not explain what exactly is going on there, or how that works.

Students who approved of the non-text modes of representation in the texts they had reviewed often referred to the additional value an illustration had, for example in helping to understand a certain technological procedure or device. The following

Table 2.3 Students' ratings of illustrations in the articles of their classmates. The total number of counts for subcategories may exceed the total number of interviewed students, since some students provided answers which had to be coded for more than one subcategory

Research question	Main category	Number of students referring to this category	Subcategory	Number of students referring to this category
What do students think of the illustrations in their classmates' articles they reviewed?	Illustrations judged negatively	41	Little connection in terms of content between illustration and text	13
			No link to illustration in text (e.g. "Fig. 2.1")	2
			Too few/no illustrations	21
			Illustrations of bad quality	3
	Illustrations judged positively	15	Illustrations help to understand content	7
			Illustrations fit the text well	8
	Layout not acceptable	6	Illustrations at wrong place (e.g. at the end of the article, instead of where the relevant information occurs in the text)	5
			Too many illustrations	1
	Other	14		
How important are illustrations in scientific texts for students?			Illustrations increase attractiveness of text	7
			Need for illustrations depends on topic	12
			Illustrations contribute to help understand the topic	49
			Illustrations are less important when the topic is well-known	4
			Illustrations attract attention	3
			Illustrations help to get started with a topic	2
			Illustrations prevent misconceptions	1
			Illustrations are generally important	9
			Too many illustrations make a text less scientific	8
			Quality of text is more important than illustrations	4
			Illustrations are helpful, if they suit the text in terms of content	4

(continued)

Table 2.3 (continued)

Research question	Main category	Number of students referring to this category	Subcategory	Number of students referring to this category
According to which criteria did students choose illustrations for their own texts?			Suggestions from Google after browsing the topic	4
			Visualisation of the possibly unknown	23
			Explanation of interrelations or technical procedures	9
			Suitability for the text	9
			Creating mental images in the reader	6
			Personal taste	6
			Making the text less dull	4
			Direct use from source literature	2
			Easiness to understand illustration	2
			Self-made, since all others were regarded as insufficient	1

example refers to an article about a dance floor which makes use of the kinetic energy of the dancers to provide electricity for lighting:

> It was about a dance floor which produces energy, and there was an illustration in it how that is working, well, when people are dancing on it what happens below. I thought this was good.

The interviews point to two important findings. First, many students were very much aware of the requirements illustrations have to fulfill, for example, that they should help the reader to understand the topic better, and that what is shown in an illustration must relate to what is described in the main text body. Secondly, analyzing each other's texts the majority of interviewees were discontent with their classmates' illustrations. From the 13 students criticizing the lack of content relation between illustration and text, eight succeeded in this respect in their own texts, two came close, and three had chosen illustrations which themselves did not seem to truly fit the text message. Consequently, many, but not all students could put into practice what they theoretically knew and criticized in others' work. This gap between theoretical knowledge and actual practice clearly needs to be worked on in school settings.

When students were asked about the significance they felt illustrations have in scientific texts, almost all of them stated that illustrations would help to understand the topic (Table 2.3). Interestingly, several students said that the need to include illustrations depends very much on topic and text. Many felt that if the text is well-written there is less need for additional information, and the same is true if the topic is well-known. Some students even felt that "too many pictures contradict the scientific nature of a text". Apparently, those students clearly differentiated between mass media articles and science papers.

Similar patterns can be found in student answers relating to the criteria they applied in illustrating their own texts. Again, most students argued that they chose a particular illustration to visualize possibly unknown or difficult to understand aspects, and several highlighted the fact that the illustrations needed to fit the text (Table 2.3).

These findings confirm the interpretation of student utterances concerning the quality and content of illustrations their fellow-students had integrated in their texts (see above). It thus seems that most of the interviewed students were very much aware of the function illustrations have in scientific and other texts, and what requirements an illustration needs to fulfill to make it a suitable one for effective communication.

Conclusion

Although writing is a rare task in many science classes, it is essential to practice scientific writing at school and thus the communication of scientific ideas and results. For the creation of interest in natural science topics, however, students might need a certain freedom in topic choice and approach. At the same time they should regard the task itself as meaningful and have the chance to develop pride about their text product. The research from a pilot study presented here shows that students' interest in natural sciences and in reading texts related to this domain may be positively influenced by article writing about self-chosen topics. Even though the demanding task to repeatedly revise their texts on a voluntary basis was not enjoyed by many and performed with varying effort and success, articles generally improved and several students explicitly stated that the revision work was helpful for them. Since interest in natural science was greater with females after the intervention, while males showed, on average, little change, this intervention seems particularly effective to create or improve females' affinity for natural science topics. Article writing and revision work in natural science subjects should thus be an integral part of the school curriculum, both for development of writing skills and for development of interest in the natural sciences. This, of course, requires corresponding training modules within pre- or in-service teacher education.

In terms of non-text representations the majority of students in the pilot project choose pictures from the internet for illustration, but they did not always use material truly aligned to the main theme of their texts. Only a minimal number of students explained illustrations with subtexts, and even fewer referred to them within the main text body. Graphs or tables were rare. First data from the follow-up project indicate that many students know about the functions illustrations serve in scientific texts (e.g. visualization of difficult content or technical procedures), and that this is partly different from, for example, journalistic pieces in which illustrations are often purely decorative with the main intention of attracting the readers' attention. However, several students did not apply this knowledge to their own texts. This awaits further research and shows the need to teach them even more intensively how to select, explain, integrate, and crosslink different representations in their written work.

For popular scientific articles such as those published in the *Young Science* magazine, student authors need to find a balance between these two extremes (scientific versus journalistic perspectives) with respect to both text and illustrations, since it is the aim of this journal to not only inform but also to raise teenagers' awareness about the fascination of natural sciences. Naturally, this is difficult for many of our authors and usually results in multiple revisions. Yet the published articles show, that all of them finally succeed. In a long-term study we plan to monitor the sustainability of this learning effect by comparing our authors written work (final theses, university assignments) with those from other students. Our hypothesis would be that both *Young Science* authors and students having participated in our program should produce better text products and are more careful in illustrating them than control group students.

Note   The *Young Science* magazine will be online from April 2015 at: http://www.young-science-magazin.com

Acknowledgements   I want to express my sincere gratitude to my colleagues from the departments of didactics of biology, chemistry and physics for their editorial help with the *Young Science* magazine, to my PhD student Sonja Enzinger and my diploma students Sandra Wallner and Thomas Mitterhuber for allowing me to include first results from their work, to the students and teachers involved in article writing and the projects presented here, to Mark Mcdermott for his many helpful suggestions to improve this paper, and to the Styrian Government (Exciting Science), the Austrian Ministry of Education (IMST) and the Dr. Heinrich-Jörg-Stiftung for their grants.

References

Ainsworth, S. (2006). DeFT: A conceptual framework for considering learning with multiple representations. *Learning and Instruction, 16*, 183–198.

Ainsworth, S., Bibby, P., & Wood, D. (2002). Examining the effects of different multiple representational systems in learning primary mathematics. *The Journal of the Learning Sciences, 11*, 25–61.

Ainsworth, S., Prain, V., & Tytler, R. (2011). Drawing to learn in science. *Science, 333*, 1096–1097.

BMUKK. (2008). Lehrplan Biologie und Umweltkunde, AHS-Oberstufe. http://www.bmukk.gv.at/medienpool/11860/lp_neu_ahs_08.pdf

Bundesagentur für Arbeit. (2012). *Arbeitsmarktberichterstattung: Der Arbeitsmarkt für Akademikerinnen und Akademiker in Deutschland – Naturwissenschaften/Informatik.* Nürnberg: Erschienen.

Christidou, V. (2011). Interest, attitudes and images related to science: Combining students' voices with the voices of school science, teachers, and popular science. *International Journal of Environmental & Science Education, 6*, 141–159.

Deci, E. L., & Ryan, R. M. (1993). Die Selbstbestimmungstheorie der Motivation und ihre Bedeutung für die Pädagogik. *Zeitschrift für Pädagogik, 39*, 223–239.

Deci, E. L., & Ryan, R. M. (2008). Facilitating optimal motivation and psychological well-being across life's domains. *Canadian Psychology, 49*, 14–23.

Donnelly, S. M. (2010). An analysis of science content and representations in introductory college physics textbooks and multimodal learning resources. Ph.D. thesis, University at Albany. http://gradworks.umi.com/3398649.pdf

Ellis, A., & Gerberich, J. R. (1947). Psychological tests and their uses. *Review of Educational Research, 17*, 64–77.

Gago, J. M., Ziman, J., Caro, P., Constantinou, C., Davies, G., Parchmannn, I., Rannikmäe, M., & Sjøberg, S. (2004). Europe needs more scientists. Increasing human resources for science and technology in Europe. Report of the high level group on human resources for science and technology in Europe. Information and Communication Unit, Directorate-General for Research, European Commission, Brussels.

Gardner, P. L. (1996). Students' interests in science and technology: Gender, age and other factors. Paper presented at the international conference on interest and gender: issues of development and change in learning, Seeon, Germany.

Gardner, P. L. (1998). The development of males' and females' interest in science and technology. In L. Hoffmann, A. Krapp, K. A. Renninger, & J. Baumert (Eds.), *Interest and learning. proceedings of the Seeon-conference on interest and gender* (pp. 41–57). Kiel: Institut fuer die Paedagogik der Naturwissenschaften (IPN).

Haider, G., & Schreiner, C. (2006). Internationaler Schülerfragebogen. https://www.bifie.at/system/files/dl/PISA-2006_fragebogen-schueler-international.pdf. 11 June 2014.

Hand, B., Hohenshell, L., & Prain, V. (2007). Examining the effect of multiple writing tasks on Year 10 biology students' understandings of cell and molecular biology concepts. *Instructional Science, 35*, 343–373.

Haste, H., Muldoon, C., Hogan, A., & Brosnan, M. (2008). If females like ethics in their science and males like gadgets, can we get science education right? Paper presented at the annual conference of the British Association for the Advancement of Science, Liverpool.

Holstermann, N., & Bögeholz, S. (2007). Interesse von Jungen und Mädchen an naturwissenschaftlichen Themen am Ende der Sekundarstufe I. *Zeitschrift für Didaktik der Naturwissenschaften, 13*, 71–86.

KMK. (2004). Einheitliche Prüfungsanforderungen in der Abiturprüfung Biologie. http://www.kmk.org/fileadmin/veroeffentlichungen_beschluesse/1989/1989_12_01-EPA-Biologie.pdf. 11 June 2014.

Krapp, A. (2005). Basic needs and development of interest and intrinsic motivational orientations. *Learning and Instruction, 15*, 381–395.

Krapp, A., & Prenzel, M. (2011). Research on interest in science: Theories, methods, and findings. *International Journal of Science Education, 33*, 27–50.

Krogh, L. B., & Thomsen, P. V. (2005). Studying students' attitudes towards science from a cultural perspective but with a quantitative methodology: Border crossing into the physics classroom. *International Journal of Science Education, 27*, 281–302.

Lachmayer, S. (2008). Entwicklung und Überprüfung eines Strukturmodells der Diagrammkompetenz für den Biologieunterricht. Ph.D. thesis, University of Kiel. http://eldiss.uni-kiel.de/macau/servlets/MCRFileNodeServlet/dissertation_derivate_00002471/Diss_Lachmayer.pdf;jsessionid=BE93C79680EA2A9C61E2F29D515EEA94?host=&o.

Lamnek, S. (2010). *Qualitative Sozialforschung*. Weinheim/Basel: Beltz Verlag.

Leisen, J. (2010). *Handbuch Sprachförderung im Fach – Sprachsensibler Fachunterricht in der Praxis*. Bonn: Varus.

Merzyn, G. (1998a). Sprache im naturwissenschaftlichen Unterricht. Teil 1. *Physik in der Schule, 36*, 203–206.

Merzyn, G. (1998b). Sprache im naturwissenschaftlichen Unterricht. Teil 2. *Physik in der Schule, 36*, 243–247.

Merzyn, G. (1998c). Sprache im naturwissenschaftlichen Unterricht. Teil 3. *Physik in der Schule, 36*, 284–287.

Merzyn, G. (2010). Kurswahlen in der gymnasialen Oberstufe. Leistungskurs Physik, Chemie, Mathematik. PhyDid B – Didaktik der Physik – Beiträge zur DPG-Frühjahrstagung.

Merzyn, G. (2013). *Naturwissenschaften, Mathematik, Technik – immer unbeliebter?* Hohengehren: Schneider Verlag.

Nitz, S. (2013). Kommunikation Schwarz auf weiß. *Unterricht Biologie, 387*(388), 34–38.

Nitz, S., Prechtl, H., & Nerdel, C. (2014). Survey of classroom use of representations: Development, field test and multilevel analysis. *Learning Environments Research, 17*, 401–422.

Norris, S. P., & Phillips, L. M. (1994). Interpreting pragmatic meaning when reading popular reports of science. *Journal of Research in Science Teaching, 31*, 947–967.

Osborne, J., Simon, S., & Tytler, R. (2009). Attitudes towards science: An update. Paper presented at the annual meeting of the American Educational Research Association, San Diego, California.

Purugganan, M., & Hewitt, J. (2004). How to read a scientific article. Cain project in engineering and professional communication. http://www.owlnet.rice.edu/~cainproj/courses/HowToReadSciArticle.pdf. Accessed 11 June 2014.

Roth, W.-M., Bowen, G. M., & McGinn, M. K. (1999). Differences in graph-related practices between high-school biology textbooks and scientific ecology journals. *Journal of Research in Science Teaching, 36*, 977–1019.

Sadoski, M. (2001). Resolving the effects of concreteness on interest, comprehension, and learning important ideas from text. *Educational Psychology Review, 13*, 263–281.

Schreiner, C. (2006). Exploring a ROSE-garden. Norwegian youth's orientations towards science – seen as signs of late modern identities. Ph.D. thesis, University of Oslo.

Statistik Austria. (2012). STATcube – Statistische Datenbank von Statistik Austria. http://statcube.at/superwebguest/autoLoad.do?db=def1509. 21 Oct 2013.

Stern, E., Aprea, C., & Ebner, H. G. (2003). Improving cross-content transfer in text processing by means of active graphical representation. *Learning and Instruction, 13*, 191–203.

Tytler, R., Prain, V., Hubber, P., & Waldrip, B. (2013). *Constructing representations to learn in science.* Rotterdam: Sense Publishers.

Vogt, H. (1998). Zusammenhang zwischen Biologieunterricht und Genese von biologieorientiertem Interesse. *Zeitschrift Zeitschrift für Didaktik der Naturwissenschaften, 4*, 13–27.

Vogt, H., Upmeier, Z. U., Belzen, A., Schröer, T., & Hoek, I. (1999). Unterrichtliche Aspekte im Fach Biologie, durch die Unterricht aus Schülersicht als interessant erachtet wird. *Zeitschrift für Didaktik der Naturwissenschaften, 5*, 75–85.

Waldrip, B., Prain, V., & Carolan, J. (2006). Learning junior secondary science through multimodal representations. *Electronic Journal of Science Education, 11*, 87–107.

Yore, L. D., & Hand, B. (2010). Epilogue: Plotting a research agenda for multiple representations, multiple modality, and multimodal representational competency. *Research in Science Education, 40*, 93–101.

Chapter 3
Effectiveness of a Lesson on Multimodal Writing

Sakari Tolppanen, Toni Rantaniitty, and Maija Aksela

Introduction

Writing-to-Learn

During the past decades, much research has been conducted on how writing could be used as a tool to learn as opposed to simply a tool to communicate already developed understanding. This research has shown that among other things, effective writing-to-learn tasks utilized in classroom settings have increased student reasoning skills, improved student ability to describe scientific concepts, and enhanced student skill in formulating arguments (Wallace et al. 2004). Therefore, writing-to-learn activities are often promoted as a way for students to make sense of new ideas (Newell 2006), as well as to help students deepen their conceptual understanding (Holliday et al. 1994) and critical thinking skills (Kieft et al. 2008).

One ongoing aim of research associated with the writing-to-learn movement has been an attempt to determine what kind of writing tasks have the greatest positive learning effects and what characterizes of writing-to-learn tasks lead to greater student learning (Yore et al. 2003). Some of these characteristics are discussed in the next section. Although some general principles have been identified, not all of them are necessarily directly transferable to science education, where often effective writing assignments consist of not only text, but also of other modes of representation, such as graphs, tables and diagrams (Lemke 1998). Effective scientific communication is typically characterized by multimodal, rather than unimodal, descriptions. Encouraging students to effectively use multimodal communication likely requires somewhat different pedagogy than the teaching methods used to

S. Tolppanen (✉) • T. Rantaniitty • M. Aksela
Department of Chemistry, The Unit of Chemistry Teacher Education,
University of Helsinki, Postbox 55, Helsinki 00014, Finland
e-mail: sakari.tolppanen@helsinki.fi

© Springer International Publishing Switzerland 2016
B. Hand et al. (eds.), *Using Multimodal Representations to Support Learning
in the Science Classroom*, DOI 10.1007/978-3-319-16450-2_3

promote unimodal products. Therefore, although, some previous research has focused on how different characteristics of scientific writing can be developed through effective teaching practices (Prain 2006), best practices associated with implementing multimodal writing tasks in science classrooms in a way that positively impacts student learning remains an area that needs to be examined.

Effective Writing Tasks

As early as the 1980s it was noted that manipulating the audience a student addresses as a part of a writing task can affect the usefulness of the writing task in terms of promoting student learning (Hillocks 1986). In a more recent study, positive effects were detected when directing a writing task to an "authentic" audience rather than to the teacher (Rijlaarsdam et al. 2006). Writing-to-learn activities, in which students are asked to write for different audiences often requires the students to translate the knowledge they possess, or find in books, into a language understandable by the audience (Gunel et al. 2009). During this translation process, students are often confronted with the inadequacy of their own understanding. This realization can initiate a process of refining and revising of the student's understanding that may not have taken place in the absence of writing to an authentic audience. However, this type of writing-task is often constructed as a part of an alternative or informal writing assignment as opposed to the creation of a more traditional lab write up. Debate persists in the literature about whether informal writing or traditional writing (reports, experiments etc.) should be used to express and potentially build students' science understanding (Wallace et al. 2004). Both writing types have been advocated as being beneficial for student learning. Informal writing, or writing using everyday language, in alternative ways and written to an audience outside of the teacher, has been promoted as a way to talk to peers about early stage scientific understanding, but the use of traditional scientific language in written communication is often argued to be more powerful in creating a more sustainable scientific understanding (Gee 2001). Ideally, combining informal and traditional writing can help students verbalize scientific language while gaining a deeper understanding of science (Wallace et al. 2004). In reference to the use of modes other than text, engaging students in creating either type of writing enhanced with the use of multiple modes would likely lead to beneficial outcomes.

A model of effective writing-to-learn activities was been developed by Prain and Hand (1996) to help teachers consider critical characteristics when developing writing-to-learn tasks for their classrooms. In this model, the authors suggest that five key elements should be considered when creating a writing task designed to effectively promote student learning. The first of these elements is *topic*. The key concepts that the writing task is dealing with must be identified. The second element to consider is *type*, which refers to specifying the type of writing task to be assigned (narrative, letter, poem etc.). A third element is *purpose*, encouraging the teacher to determine the role that the task will fill in the process of concept development (review, clarify, apply etc.). The fourth element to consider is *audience*, referring to whom the written task is directed to (peers, teacher, younger student, etc.) and the last element is *method of text production*, referring to how the students will physically create the text (computer, handwritten, groups, etc.). Consideration and manipulation of these factors will not only allow teachers to develop a variety of writing-to-learn tasks, but it will help to assure that the tasks go beyond repeating of already learned information and encourage students to develop new understanding as a part of the writing task itself.

In a more recent study, Prain and Hand (2005) also suggest that in effective writing activities, students are engaged in translating scientific ideas in three different ways. The first translation takes place when the students must translate the targeted concepts for themselves so that they can understand the concept well enough to communicate to others about it. The second translation involves a process of changing from the way the concept has been described to or by the student into an appropriate form to communicate to the intended audience. For example, a secondary student writing to an elementary student would need to develop a method of accurately communicating a science concept using vocabulary appropriate for their audience and this vocabulary would likely be different than the vocabulary they encountered in their original engagement with the concept. Understanding of the audience will influence how well this translation is carried out. The third translation happens when the targeted concepts have to be translated back into appropriate scientific vocabulary, when completing classroom assessments.

Multimodal Representations in Writing

Lemke (1998) has stated that science is not done and communicated with verbal language alone, but rather, combines verbal text with mathematical expressions, quantitative graphs, information tables, abstract diagrams, maps, drawings, photographs and requires a combination of visual tools seen nowhere else. In other words, science is communicated through multimodal representations. For students to understand science, it is therefore important that they are able to understand and use multimodal representations. Furthermore, teaching students to write using multimodal representations will likely help them build their conceptual understanding in science, as well as reap the other, previously mentioned, benefits of effective writing-to-learn tasks (Bangert-Drowns et al. 2004; Prain 2006).

When scientific knowledge is presented to students through different modes, students need to integrate the pieces of knowledge in order to form a more complete conceptual understanding of the topic (Airey and Linder 2006). According to Seufert (2003), students need to form referential connections between different modes, in order to progress through a more meaningful learning sequence. Likewise, diSessa (2004) notes that the chances of obtaining a deeper understanding of a topic are increased when different modes of representation are introduced. He also emphasizes that it is important that students not only learn to use and read different modes of interpretation, but also learn about the limitations of different modes of representation including how they may be interpreted incorrectly.

The use of multimodal writing has also been tested in classroom settings. For instance, in research conducted by Hand et al. (2009), students were given different kinds of writing tasks, some of which required the use of multimodal representations and others that did not. The post-test results indicated that students who use mathematical equations in addition to text in their writing products in physics, constructed greater understanding the concepts. In addition, using other alternative modes, such as images, has been shown to support learning more than the use of text-only products, but to a lesser degree than when mathematics equations were used. In another study McDermott and Hand (2013) provided a test group with a lesson on the use of multimodal writing. Test students were then asked to create a multimodal product. Test group products were compared to a control group who were given the same task without the lesson on multimodal writing. After completing the written assignments, students were tested on content knowledge and students from the test group outperformed the control group in several areas.

However, research has not consistently supported the benefits of writing-to-learn, or provided a clear indication of the benefit of using multimodal writing (Galbraith and Torrance 1999; Holliday et al. 1994; Klein 1999; Rivard 1994; Yore et al. 2003). For instance, Nieswandt and Bellomo (2009) found that writing is not a completely suitable method to measure student learning in different contexts. Furthermore, while some research has shown writing tasks increase student learning and metacognitive skill (Boscolo and Mason 2001; Hand et al. 2004), others have shown mixed results (Ackerman 1993), and in a meta-analysis, Bangert-Drowns et al. (2004) showed that positive effects are not common and effect sizes are small.

Therefore, best practices associated with writing-to-learn tasks and the use of multiple modes of representation are not yet clear. One first step in clarifying this situation is to explore how students use multimodal representations to present their ideas. We concur with Wallace et al. (2004) that students learn to clarify their thoughts when seeking the appropriate form in which to present their ideas to an audience. Therefore the research discussed here seeks to begin to explore student learning from multimodal tasks by focusing on how students choose to represent their ideas multimodally. In part, we aim to find out whether there is a difference in the use of modes between females and males. In a study by Tarmo (1991), it was argued that females in Finland do not understand science concepts as well as males, often finding them too abstract. However, this argument is not in accord with the PISA results from Finland, where gender differences were non-existent in science knowledge (Lavonen and Laaksonen 2009). Research does, however, show that females, more easily than males, perceive their understanding as lacking (Airey and Linder 2006).

Aim

The aim of this research is to answer the following questions:

1. How does a lesson on multimodal writing effect students' text production and use of alternative modes of representation in a science writing task?
2. Is there a gender difference in the use of multimodal writing?
3. How do different lesson structures affect students' ability to use multimodal writing?

Based on these results, we will discuss whether a single lesson on multimodal writing helps students assess their own writing and how the use of multimodal writing tasks can be implemented into the classroom.

Method

Participants

This study was carried out in three randomly selected schools in a large city in, Finland. A total of 98 grade 8 students participated, including 53 males and 45 females. Fifty-four students were in the test group and 44 in the control group.

School A was a regular Finnish comprehensive school, consisting of grades 7–9. From school A, 41 students took part in the study, of which 27 were part of the test group. The teacher giving the lesson in school A had 6 years of teaching experience. She gave the lesson on multimodal writing to two student groups (N = 13 and N = 14). The control-group was given a regular lesson by the teacher and that class consisted of 14 students. In this regular lesson, students read the chapter, took notes written by the teacher and did some hands-on experiments.

School B has grade 7 through grade 12 students, with a comprehensive school and an upper secondary school in the same building. The upper secondary school (grades 10–12) offers a special science program. The students in this study were in grade 8 and therefore were not a part of the science program. However, the teacher teaching the class also teaches courses in the science program. In school B, 43 students took part in this study, with 27 in the test group and 17 in the control group. The teacher in school B had 5 years of teaching experience and she gave the lesson on multimodal writing to two student groups (N = 12 and N = 15).

School C was also a comprehensive school. In school C, the teacher was unable to give the lesson on multimodal writing due to changes in school schedules, so data was collected only from the control group of 14 students. The teacher in school C had 12 years of teaching experience.

Lesson on Multimodal Writing

A single lesson approach, based on research by McDermott and Hand (2013) was used in this research. Two weeks before the scheduled lesson, researchers met up with each teacher to discuss that the aim of the research was to examine the effects of a single lesson about multimodal writing on students' ability to use alternative modes of representation. Teachers were then given guidance on how to give a lesson on multimodal writing by discussing the following step-by-step instructions:

1. Provide students a text description of a challenging topic closely aligned to a current concept being dealt with in the class. The text should not have any modes of representation other than text.
2. Ask students to analyze the text to determine how effective the text is in communicating the scientific information.
3. Ask students to look at different science sources and analyze how these sources use alternative modes of representation to support written communication (tables, graphs, image etc.).
4. Based on their observations, ask students to generate a list of the alternative modes used in common science sources (text book, magazine etc.) and strategies utilized to link the modes to text.
5. Based on their list, guide students in generating an assessment matrix that can then be used to evaluate the use of alternative modes of representation in any written product.
6. Practice using the matrix to analyze a scientific source with students.
7. Give students homework where they need to practice using the matrix to analyze a specific piece of scientific writing.
8. Refer to multimodal writing in the following lessons.
9. After 2 weeks, give the students an assignment, where they need to write about a concept dealt with in class. Give instructions to use both written text and alternative modes of representation when explaining the issue.

During the multi-modal lesson, teachers were asked to keep track of how much time they used for each section of the lesson in a table provided to them.

Each of the three schools taught a different topic in science during the testing procedures, therefore, they used a different initial text (1), analyzed different sources (3) and gave different homework tasks (7) relating to the particular theme they were studying. As homework, teachers asked the students to create a multimodal product where they were required to use alternative modes of representation. Students were instructed to do the assignment using the guidelines of the matrix created. In school A, the students were asked to explain how electricity could be produced on a deserted island by using some specified scraps from an air-plane. In school B, students were asked to explain how it is possible that with one liter of gasoline a car travels further than with one liter of ethanol.

While the test group had the lesson on multimodal writing, the control group was taught a traditional lesson on the same science theme that the multimodal writing lesson dealt with. Both groups used the same amount of class time to deal with the science concept and both groups were given the same homework, with the only difference being that the test group was asked to use alternative modes in presenting their homework.

Writing Assignment

Two weeks after the lesson on multimodal writing, the students were given an assignment in which they were told that they were the editor of a science magazine and needed to provide an answer to one of the questions asked by a reader. The assignment was related to the topic they had been studying in the previous class. Students in school A wrote about fats and carbohydrates, students from school B wrote about electricity and students from school C wrote about the expansion of metals.

The assignments were created so that they considered the model for an effective writing-to-learn activity (Prain and Hand 1996), namely students were asked to write to some-one who did not yet understand a concept that they understood themselves. For example, the assignment given to school C was as follows:

> You are the editor of a science magazine and you are given the task to answer a question asked by the reader.

> Question: Years ago when I was traveling, I noticed that on a hot summer day the railway tracks were so bent that the railway traffic needed to be stopped. However, it did not seem like the tracks were bent by human force. Is there a scientific explanation to this?

In the task, students were asked to use multimodal writing to communicate their ideas. The students had one week to complete the assignment as homework with the assignments were then evaluated by their teacher. The control group and the test group were given the same assignment and the same instructions to complete the task, except that the students in the test group were asked to reflect on the earlier lesson and notes on multimodal writing.

Data Analysis

In this research, the text production and the use of alternative modes was analyzed utilizing a rubric based on a similar rubric described by McDermott and Hand (2013). The rubric was developed by consulting with international researchers who have been involved in multimodal writing research in the past and previous research on the subject (McDermott and Hand 2013). For this research, small adjustments were made to the rubric to better fit the type of products the students were asked to complete.

Using this rubric, the data collected from the students was evaluated on three different characteristics: *text production*, *general mode-analysis*, and *individual mode-analysis*. The text production examined the use of text only in explaining ideas. The general and individual mode analysis examined how other alternative modes of representation, such as images, tables and graphs, were used to explain an idea. Each of these three sections contained sub-categories that were given a score between 0 to 2, using the guidelines explained below. Discussion among researchers on how to interpret these guidelines was used to attempt to standardize assessment. Ten percent of the assignments were then randomly selected and analyzed by two researchers. Their analyses were compared and the small differences that were found were discussed to form an even clearer assessment. After giving the scores in *text production*, *general mode-analysis* and *individual mode-analysis*, the data was analyzed using the Mann–Whitney U-test as the data was small and not normally distributed.

Part I: Text Production

The *Text production* assessment was comprised of analyzing both *Assignment Expectations* and *Audience Consideration*. In *Assignment Expectations*, the writing process was analysed with the following characteristics:

- Grammar: Students wrote the text according to grammar rules and avoided misspelling.
- Coverage of Required Topics: Students presented the main concepts of the topic required by the assignment.
- Accuracy of Science Concepts: Students accurately described the science content discussed in the written product.
- Completeness of Meaning: Students thoroughly covered all aspects of required topics.
- Logical Order of Text: The chapters/paragraphs were in a logical and constructive order.

In *Audience Considerations*, the suitability for the audience was analysed based on the following characteristics:

- Appropriate Language/Vocabulary: Students used language appropriate for the audience of a typical science magazine.
- Identification of Key Terms (Underlined, Highlighted, Italics): Students emphasized the key terms associated with the topic by underlining, highlighting and the appropriate use of italics.
- Headings: Students used logical headings, and started writing with a form that addresses the reader (e.g. Dear, Sir/Madam…)

For each of the five characteristics in *Assignment Expectations*, and for the *appropriate language/vocabulary* characteristic in *Audience Consideration*, a score of zero, one or two points was given, depending on how expectations were met.

Zero points were given if no evidence of the characteristic was presented, one point was given when there was some evidence that the characteristic was present and two points when the characteristic was applied throughout the task. The characteristics of *Identified Key Terms* and *Headings* in *Audience Consideration* were evaluated with a two-point scale, where 0 points was given if a characteristic was not employed at all, and 1 point was given if the characteristic was utilized. Therefore, in *text production*, a student could attain a maximum score of 14 points.

Part II: General Alternative Mode Analysis

The *General Alternative Mode Analysis* focused on the different modes students utilized throughout the written product and how appropriate they were for the task. In this part of the analysis, the following characteristics were analysed:

- Modes Appropriate for Audience: The modes were classified as appropriate when students took into account the audience and their preliminary knowledge of the topic.
- Key Terms Included in Modes: The key terms appeared in modes, which helped the audience understand what specific information the modes presented.
- Accurate/Correct Representations: The content included in the modes was scientifically accurate/correct.
- Modes Linked to Main Concepts: The context of the modes was related to the topic or linked to main concepts.
- Logical Distribution of Modes: The student placed the modes in a logical order in respect to the written text. In other words, the relevant mode was placed next to, or below the part of the text where that particular concept was being discussed. Points were deducted if a mode was placed at the bottom of the page (or another page), though it was related to the text at the top of the page.

As in the previous section, the characteristics were given a score of zero, one or two depending on how well they were implemented. The maximum amount of points from *General Alternative Modes Analysis* was therefore 10 points.

Part III: Individual Alternative Mode Analysis

In the third part of the analysis, each mode was studied individually. The evaluation of the individual modes focused on (A) *Embeddedness Strategies* and (B) *Characteristics. Embeddedness Strategies* evaluate how students linked their modes to text using the following criteria:

- Caption: Students used captions to explain the content of the modes.
- Next to Appropriate Text: Students' modes had the same content or dealt with the same content as the text close to them.

- Referred to in Text: Students referenced their modes in text, for example "As the figure above showed…"
- Explained in Text: Students explained the modes or the content of the modes in text.

Characteristics was analysed by including the following criteria for the evaluation:

- Scientifically Accurate: The individual mode was scientifically accurate.
- Necessary for Explanation: Modes and captions were understandable together, helping the audience understand what topics were presented.
- Conceptual Connection to Text: The concepts presented in the individual modes were relevant to the topic of the written text.

In *Embeddedness Strategies*, a score of 0 (not employed) or 1 (utilized) was given based on the quality of the mode. In *Characteristics*, a score of 0 (no attempt), 1 (limited attempt) or 2 (entire/complete) was given, based on the quality of the mode. The maximum amount of points attained from *Individual Alternative Mode Analysis* was therefore 10 points.

Results

This section will first look at what kind of effect a single lesson on multimodal writing has on text production and the use of alternative modes, the effect of gender on the use of multimodal writing, and finally, the effects of the differences in lesson plans.

The Effects of a Lesson on Multimodal Writing on Text Production and Use of Alternative Modes

When the test group and the control group were compared with each other, differences were detected in the way students used text and alternative modes of representation. When analyzing the use of *Text Production*, of the test group (n=54), 8 students did not use text in their assignment. Without written text, the message of the student's letter was unclear, and therefore these results were omitted from the analysis. When analyzing the remaining texts (n=90), the test group (M=0.46, SD=0.504) outperformed (Z=−2.521, p<0.05) the control group (M=0.20, SD=0.408) in the use of headings. As seen from the mean value, this was mainly due to the fact that the control group did not use headings at all. Significant differences were not found in the other characteristics of *Text Production* between the two groups.

When observing the use of *Alternative Modes*, analysis showed that 20 % (n = 11) of the test group and 43 % (n = 19) of the control group did not use alternative modes of representation to support their text. A total of 68 students used alternative modes in addition to text. The students used a total of 104 alternative modes, 67 by the test group and 37 by the control group. Among the students who did use alternative modes of representation, students from the test group used an average of 1.56 modes and the control group used an average of 1.48 modes in their assignment. When those students not using alternative modes (n = 30) were omitted, the control group (M = 1.32, SD = 0.476) outperformed (Z = −2.810, p < 0.01) the test group (M = 0.86, SD = 0.675) in *Logical Distribution* of *General Alternative Modes*. In the individual alternative modes, the control group (M = 0.19, SD = 0.397) outperformed (Z = −2.352, p < 0.05) the test group (M = 0.05, SD = 0.210) in *referring to their modes in text*.

The Effect of Gender on the Use of Multimodal Writing

Differences in *text production* and alternative modes of representations were detected between females and males. In *text production*, significant differences between genders were seen in most of the characteristics assessed. When both the test group and control group were analyzed together (N = 68), females outperformed males in *Grammar* (Z = −2.110, p < 0.05), *Accuracy of Science Concepts* (Z = −2.097, p < 0.05) *Completeness of Meaning* (Z = −2.409, p < 0.05) and *Appropriate Language/Vocabulary* (Z = −3.785, p < 0.001). As a result, when all of the characteristics of *Text Production* were combined to one overall total value, females significantly outperformed males (Z = −3.340, p < 0.01). See Table 3.1 for mean and SD values.

When genders were compared in the test-group (N = 46) only, females outperformed males in *Coverage of Required Topic* (Z = −2.772, p < 0.01), *Completeness of Meaning* (Z = −2.528, p < 0.05) and *Appropriate Language/Vocabulary* (Z = −4.611, p < 0.001). As a result, when all of the characteristics of *Text Production* were combined to one overall total value, females significantly outperformed males

Table 3.1 Gender differences in text production

Gender		Grammer	Accuracy of scientific concepts	Completeness of meaning	Appropriate language	All "text production" combined
Male (N=45)	Mean	1.44	1.07	1.53	1.24	8.18
	SD	.503	.654	.661	.609	2.790
Female (N=45)	Mean	1.67	1.36	1.82	1.71	9.91
	SD	.477	.645	.442	.506	2.448
Total (N=90)	Mean	1.56	1.21	1.68	1.48	9.04
	SD	.500	.662	.577	.604	2.752

Table 3.2 Gender differences in text production within test group

Gender		Coverage of required topics	Completeness of meaning	Appropriate language	All "text production" combined
Male	**Mean**	1.00	1.41	1.15	7.81
(N=27)	**SD**	.620	.747	.662	3.039
Female	**Mean**	1.53	1.89	2.00	10.84
(N=19)	**SD**	.513	.315	.000	1.463
Total	**Mean**	1.22	1.61	1.50	9.07
(N=46)	**SD**	.629	.649	.658	2.909

Table 3.3 General alternative modes of test group

Gender		Modes linked to main concepts	Logical distribution of modes
Male (N=27)	**Mean**	1.74	.70
	SD	.447	.669
Female (N=16)	**Mean**	2.00	1.13
	SD	.000	.619
Total (N=43)	**Mean**	1.84	.86
	SD	.374	.675

($Z=-3.674$, $p<0.001$) (see Table 3.2). When gender differences were tested in the control group (N=44) significant differences between males and females were not noted.

Gender differences were also noted in the use of alternative modes of representation. When both the test-group and control-group were analyzed simultaneously in the use of *General Alternative Modes*, females outperformed males in *Modes Linked to Main Concepts* ($Z=-2.537$, $p<0.05$) as well as *Logical Distribution of Modes* ($Z=-2.687$, $p<0.01$). The same differences were found when only analyzing gender differences in the test group ($Z=-2.200$, $p<0.05$ and $Z=-2.003$, $p<0.05$ respectively) (See Table 3.3). However, analysis of only the control group was not possible, as the group was too small after zero values were omitted.

When comparing the *Individual Alternative Modes*, females outperformed males in *Explaining their Modes in Text* ($Z=-3.271$, $p<0.01$), *Scientific Accuracy* ($Z=-2.154$, $p<0.05$) and in making *Conceptual Connections to Text* ($Z=-2.527$, $p<0.05$). Furthermore, the *Individual Mode Analysis* showed that when all the scores for the individual characteristics are added together in the category, females outperformed ($Z=-2.578$, $p=0.01$) males significantly. When only the test-group was examined, females outperformed males in *Location* ($Z=-2.352$, $p<0.05$) and in *Explanation* ($Z=-3.671$, $p<0.001$). Furthermore, the *Individual Mode Analysis* showed that when all the characteristics within the group are analyzed simultaneously, females significantly outperformed ($Z=-2.170$, $p<0.05$) males. See Table 3.4 for mean and SD values.

Table 3.4 Gender differences in individual alternative modes within test group

Gender		Next to appropriate text	Explained in text	All "individual alternative modes" combined
Male (N=37)	Mean	.43	.46	5.78
	SD	.502	.505	2.002
Female (N=29)	Mean	.72	.90	6.83
	SD	.455	.310	1.416
Total (N=66)	Mean	.56	.65	6.24
	SD	.500	.480	1.832

For males (n=53), fifteen (28 %) did not use multiple modes of representation and eight did not use text in their assignments. For females (n=45), fifteen (33 %) did not use alternative modes of representation, but all of them used text. From those who used alternative modes, the females used an average of 1.77, and males an average of 1.34 modes.

The Effect of Lesson Plan on the Use of Multimodal Writing

Some differences between the lessons taught on multimodal writing were observed. In relation to overall time, the teacher from school A used 57 min to complete the lesson on multimodal writing, but the teacher from school B used just 40 min, as they decided to leave out steps 6 and 7. In relation to time allocation for different aspects of the lesson, in school B, slightly more time (10 min vs. 8 min) was given to analyze the text-only description provided to the students at the beginning of the class. Also, in school B, 50 % more time (15 min vs. 10 min) was spent looking at how different modes of representation were used in different sources. When analyzing the sources, students in school B only analyzed their own text-book, whereas students in school A looked at their textbook and Wikipedia. Since schools have the option of choosing their own textbooks, the two schools were using different textbooks. A textbook analysis showed that the textbook of school A used different modes of representation more frequently than the book used at school B. However, in the book used by school A, modes were missing captions and explanations, and some images were not related to the topic being presented, but seemed to be placed there mainly for looks. In the book used by school B, different modes of representation were strategically placed in the text, explanations of modes were always present and images were used in a way that supports the content of the text. In both of the schools, the students made a list of the different modes that they found in the science sources and created a matrix out of them. However, only students in school A practiced using this matrix on a new scientific source to evaluate the use of modes in that text.

After the lesson, homework was assigned in both schools. In school A, the homework was to practice the use of the matrix as an evaluation tool by applying it to a student's own written assignment, whereas in school B, students were instructed to pay close attention to how they use different modes of representation in a written

Table 3.5 School differences in the use of alternative modes within test group

School		Next to appropriate text	Explained in text	Scientifically accurate	Conceptual connections to text	All "alternative modes" combined
A (N=37)	Mean	.38	.49	1.11	1.68	5.49
	SD	.492	.507	.567	.475	1.820
B (N=29)	Mean	.79	.86	1.76	1.93	7.21
	SD	.412	.351	.435	.258	1.346
Total (N=66)	Mean	.56	.65	1.39	1.79	6.24
	SD	.500	.480	.605	.412	1.832

Table 3.6 School differences in the use alternative modes within control group

School		Caption	Necessary for explanation	Conceptual connections to text	All "individual alternative modes" combined
A (N=14)	Mean	.57	1.43	2.00	7.00
	SD	.514	.514	.000	1.359
B (N=10)	Mean	.00	1.00	1.50	5.80
	SD	.000	.000	.527	1.033
Total (N=24)	Mean	.33	1.25	1.79	6.50
	SD	.482	.442	.415	1.351

task and were asked to reflect this in the matrix created. However, they were not asked to use the matrix to evaluate their own work.

Due to the large amount of students who did not use either written text or alternative modes, comparing the test group and control group in *Text Production* and *General Alternative Modes* across schools was impossible due to small sample size. However, analysis on the individual alternative modes was possible. In the test group, school B outperformed school A in Location ($Z=-3.344$, $p=0.001$), Explanation ($Z=-3.154$, $p<0.01$) Accuracy ($Z=-4.428$, $p<0.01$), Connection to Text ($Z=-2.499$, $p<0.05$) and total score from all categories ($Z=-3.861$, $p<0.001$). See Table 3.5 for mean and SD values.

When comparing the control groups across schools, the control group at school A outperformed school C in Captions ($Z=-2.866$, $p<0.01$), Necessary for Explanation ($Z=-2.340$, $p<0.05$) and Conceptual Connection to Text ($Z=-2.911$, $p<0.01$). Also, when all of the scores from all sections in Individual Alternative Modes were added together, a significant difference was noted ($Z=-2.039$, $p<0.05$) favoring school A over school C. However, in the control group the sample size was so small (N=24) that the data cannot be considered representative (Table 3.6).

Discussion

Difference Between Test Group and Control Group

This research gives insight on the effectiveness of a single multimodal writing lesson. When comparing the test group and control group, it is seen that a single lesson on multimodal writing did not appear to have an impact on students' *Text Production*. This is a logical result, as all of the characteristics assessed required a clear understanding of general writing structure and that was not the main focus of the multimodal lesson. In addition, it is likely most students have had similar prior instruction related to general writing ability. One difference between the test and control group that was detected was the use of *headings*, indicating the use of headings may be a strategy that is easily implemented into writing once students are made aware of it. Therefore, the single multimodal lesson gave students some idea of how to improve their multimodal product and easy-to-implement strategies were utilized, but more in-depth lessons would be required to encourage more complex strategy use and the production of more fully integrated multimodal products.

One unanticipated consequence of participation in the multimodal writing lesson was that 8 students from the test group did not utilize any text at all in their written product. One potential reason for this could be that students tend to do what they expect will be assessed (Snyder 1973) and perhaps the emphasis on alternative modes in the multimodal lesson led students to a belief that the use of modes other than text was of more critical importance than the integration of text with other modes. Asking students to analyse and give feedback on each other's work before handing it in could prevent this problem from reoccurring. Although the *text production* did not seem to be greatly impacted by the multimodal lesson, a clear difference in use of alternative modes did develop between the test and control groups, as 80 % of the test group compared to only 57 % of the control group used them to support their writing.

Interestingly, after a single multimodal lesson, the increased use of alternative modes did not increase the overall quality of the written products. In fact, the control group performed better in both the *Distribution of Modes*, as well as *Referring to Modes* in their text. One possible explanation, supported by the increased use of alternative modes, is that the message the students took from the multimodal lesson was the importance of using a number of different modes rather than the importance of effectively integrating different modes with text. Based on the low percentage of use of alternative modes in the control group, perhaps only those who were previously familiar, or confident in using alternative modes, did so and these students may have been more likely to use strategies to effectively integrate the different modes.

These findings seems to indicate that the lesson on multimodal writing, while encouraging the use of more modes, did not encourage the more effective use of modes and the effective integration of modes with text. It is likely that this effective integration is what is needed to better understand a concept, as suggested by Seufert (2003). Furthermore, if students better understood the effective use of multimodal writing, it could be used as a way of making sense of new ideas (Newell 2006). However, the fact that the students scored relatively low on scientific accuracy in both text and in their alternative modes indicates this sense making was not encouraged, possibly due to the lack of effective integration of text and modes. Therefore, further research on how to emphasize the importance of effectively combining text and alternative modes as the goal of multimodal writing needs to be conducted. Implementing multimodal writing into several lessons could help students understand the role of the alternative modes, but also, lesson plans should be modified to examine which teaching techniques work best to promote this.

The Effect of Gender on the Use of Multimodal Writing

The fact that gender differences did occur in the test group, but not in the control group, may support a claim that females understand the importance of integrating written text with other modes in multimodal writing better than males. Earlier research (Välijärvi et al. 2007) has shown that in Finland there is a large gap in reading skills between males and females, and this research implies that this gap may also transfer into writing, as others (Fitzgerald and Shanahan 2000) have shown that a correlation between reading and writing skills often exists. In this regard, if females tend to be more advanced in their writing skills in general, they may be more likely to move from the simple use of alternative modes with text to effective integration of alternative modes with text more quickly.

This research supports the contention that higher writing capability may be related to the ability to use alternative modes of representation more accurately. This relationship is especially seen in the level of *Individual Alternative Modes*, where females outperformed males in several assessed categories. Besides the possibility that writing capability is related to the ability to use alternative modes, it is also possible that females are better able to understand their audience by realizing that the readers' understanding of the targeted topic may be lacking. According to a recent study (Airey and Linder 2006), females perceive their own understanding to be lacking more easily than males and this could transfer to how the females see the audience. This idea is supported by the fact that females tend to use more modes in their writing than males (mean values of 1.77 and 1.34 respectively), and that they view this as a way to improve communication to an audience with a perceived lack of understanding.

It is important to consider that the gender differences noted could also be impacted by the interest in the topics given. Some students may find a specific writing topic uninteresting or difficult, causing them to be less motivated about the given task, as seen in previous research (Gee 2004; Räsänen 1991; Tarmo 1991). Interest could also play a role in reading and writing, as discussed in a research done by The Finnish National Board of Education (Lappalainen 2011).

The Effect of the Lesson Plan on the Use of Multimodal Writing

Differences were noted in the actual implementation of the multimodal lesson, mainly in the amount of time used in different sections of the lesson. Although the relatively small size of the test groups precludes a thorough analysis of potential differences due to differences in implementation of the lesson, results did show a clear difference in the schools, with school B outperforming school A in several of the categories. This finding at least begins to support the claim that not only the presence or absence of a lesson on multimodal writing can impact student performance, but that the specific make-up of the lesson can have an effect.

School A used more time on the lesson on multimodal writing, and students had a chance to practice using their matrix, which may have led to a greater impact on students. Secondly, school B is connected to a science-oriented upper secondary school. The teachers therefore teach in both the junior high and the senior high, so their expectations and teaching methods may differ from the teachers of other schools. An example of this was seen in how the teachers in school B decided to omit the practice of the matrix and gave the students a more open-ended homework assignment. Their vast experience with different kinds of students has the potential to increase their efficacy, a trait that is seen as one of the most important teacher traits to increase student motivation and learning (Ashton 1984).

Another explanation for differences between student outcomes in different schools could be the textbooks used in the schools. School B used a text book that, though it used fewer alternative modes of representation than the book used by school A, used them as a tool to help the reader better understand the text.. For instance, in the book used by school B, the alternative modes of representation were more likely to have descriptive captions and were more regularly referred to in the text, than in the book used by school A. Furthermore, in the book used by school A, alternative modes of representation tended to be irrelevant to the text, bringing colour to the pages, but not enhancing learning. As the text book is one of the primary sources students utilize, they are most likely influenced by the way things are represented in the textbook.

Conclusions

This research found that a single lesson on multimodal writing increases the amount of multimodal representations used by students taking part in the lesson. However, it also suggests that teaching a single lesson on multimodal writing is not sufficient to increase the quality of the alternative modes used and to encourage students to more effectively integrate alternative modes with text. From these findings, we conclude that a single lesson on multimodal writing will begin to help students understand the importance of multimodal writing, but is not enough for students to fully develop an understanding of how alternative modes should be integrated in a way that leads to deeper student understanding. Therefore, further research is needed on how to best implement multimodal writing lessons into school.

This research also implies a potential link between students writing skills and the use of alternative modes exists. Further research on this link is required to determine how to more effectively design instruction that encourages effective multimodal writing for students of differing writing ability levels.

Note This article is a modified version of the article published in *LUMAT: Research and Practice in Math, Science and Technology Education.* (Vol 1, No. 5, 2013)

Acknowledgements We would like to thank the teachers who gave their valuable time in order to make this study possible.

References

Ackerman, J. M. (1993). The promise of writing to learn. *Written Communication, 10*(3), 334–370.

Airey, J., & Linder, C. (2006). *Languages, modality and disciplinary knowledge.* Unpublished manuscript from http://www.slidefinder.net/m/multimod/multimod/15642224

Ashton, P. (1984). Teacher efficacy: A motivational paradigm for effective teacher education. *Journal of Teacher Education, 35*(5), 28–32.

Bangert-Drowns, R., Hurley, M. M., & Wilkinson, B. (2004). The effects of school-based writing-to-learn interventions on academic achievement: A meta-analysis. *Review of Educational Research, 74*(1), 29–58.

Boscolo, P., & Mason, L. (2001). Writing to learn, writing to transfer. In L. M. Tynjala & K. Lonka (Eds.), *Writing as a learning tool: Integrating theory and practice* (pp. 83–104). The Netherlands: Kluwer.

diSessa, A. A. (2004). Metarepresentation: Native competence and targets for instruction. *Cognition and Instruction, 22*(3), 293–331.

Fitzgerald, J., & Shanahan, T. (2000). Reading and writing relations and their development. *Educational Psychologist, 35*(1), 39–50.

Galbraith, D., & Torrance, M. (1999). Conceptual processes in writing: From problem-solving to text production. In D. Galbraith & M. Torrance (Eds.), *Knowing what to write: Conceptual processes in text production* (Studies in writing, pp. 1–12). Amsterdam: Amsterdam University Press.

Gee, J. P. (2001). Reading as situated language: A sociocognitive perspective. *Journal of Adolescent and Adult Literacy, 44*(8), 714–725.

Gee, J. P. (2004). Language in the science classroom: Academic social languages as the heart of school-based literacy. In W. Saul (Ed.), *Crossing borders in literacy and science instruction: Perspectives in theory and practice* (pp. 13–32). Newark: International Reading Association/National Science Teachers Association.

Gunel, M., Hand, B., & McDermott, M. A. (2009). Writing for different audiences: Effects on high-school students' conceptual understanding of biology. *Learning and Instruction, 19*(4), 354–367.

Hand, B., Wallace, C. W., & Yang, E. (2004). Using a science writing heuristic to enhance learning outcomes from laboratory activities in seventh-grade science: Quantitative and qualitative aspects. *International Journal of Science Education, 26*(2), 131–149.

Hand, B. M., Gunel, M., & Ulu, C. (2009). Sequencing embedded multimodal representations in a writing to learn approach to the teaching of electricity. *Journal of Research in Science Teaching, 46*(3), 225–247.

Hillocks, G. (1986). *Research on written composition: New directions for teaching.* Urbana: ERIC Clearinghouse on Reading and Communication Skills and National Conference on Research in English.

Holliday, W. G., Yore, L. D., & Alvermann, D. E. (1994). The reading-science learning-writing connection: Breakthroughs, barriers, and promises. *Journal of Research in Science Teaching, 31*(9), 877–893.

McDermott, M. A., & Hand, B. (2013). The impact of embedding multiple modes of representation within writing tasks on high school students' chemistry understanding. *Instructional Science, 41*(1), 217–246.

Kieft, M., Rijlaarsdam, G., & van den Bergh, H. (2008). An aptitude–treatment interaction approach to writing-to-learn. *Learning and Instruction, 18*(4), 379–390.

Klein, P. (1999). Reopening inquiry into cognitive processes in writing-to-learn. *Educational Psychology Review, 11*(3), 203–270.

Lappalainen, H. (2011). *Sen edestään löytää – äidinkielen ja kirjallisuuden oppimistulokset perusopetuksen päättövaiheessa 2010.* [What goes around, comes around -final learning achievements in Finnish language of students ending middle-school in 2010]. Helsinki: Opetushallitus/ Finnish National Board of Education.

Lavonen, J., & Laaksonen, S. (2009). Context of teaching and learning school science in Finland: Reflections on PISA 2006 results. *Journal of Research in Science Teaching, 46*(8), 922–944.

Lemke, J. (1998). Multiplying meaning: Visual and verbal semiotics in scientific text. In J. R. Martin & R. Vccl (Eds.), *Reading science: Critical and functional perspectives on discourses of science* (pp. 87–113). London: Routledge.

Newell, G. E. (2006). Writing to learn: How alternative theories of school writing account for student performance. In C. MacArthur, S. Graham, & J. Fitzgerald (Eds.), *Handbook of writing research* (pp. 235–247). Guilford: The Guilford Press.

Nieswandt, M., & Bellomo, K. (2009). Written extended-response questions as classroom assessment tools for meaningful understanding of evolutionary theory. *Journal of Research in Science Teaching, 46*(3), 333–356.

Prain, V. (2006). Learning from writing in secondary science: Some theoretical and practical implications. *International Journal of Science Education, 28*(2), 179–201.

Prain, V., & Hand, B. (1996). Writing for learning in secondary science: Rethinking practices. *Teaching and Teacher Education, 12*(6), 609–626.

Prain, V., & Hand, B. (2005). Science and literacy. In K. Appleton (Ed.), *Elementary science teacher education: Issues and practice* (pp. 154–174). Mahwah: Association of Educators of Science Teachers Publication.

Räsänen, L. (1991). Tytöt ja fysikaalisten käsitteiden oppiminen [Girls and concept learning in physics]. *Kasvatus, 22*(3), 185–194.

Rijlaarsdam, G., Couzijn, M., Janssen, T., Braaksma, M., & Kieft, M. (2006). Writing experiment manuals in science education: The impact of writing, genre, and audience. *International Journal of Science Education, 28*(2–3), 203–233.

Rivard, L. O. P. (1994). A review of writing to learn in science: Implications for practice and research. *Journal of Research in Science Teaching, 31*(9), 969–983.

Seufert, T. (2003). Supporting coherence formation in learning from multiple representations. *Learning and Instruction, 13*(2), 227–237.

Snyder, B. (1973). *The hidden curriculum.* New York: Knopf.

Tarmo, M. (1991). Opettajan sukupuolilinssit [Teachers' gender-glasses]. *Kasvatus, 22*(3), 195–204.

Välijärvi, J., Kupari, P., Linnakylä, P., Reinikainen, P., Sulkunen, S., Törnroos, J., & Arfmann, I. (2007). In J. Välijärvi, P. Kupari, P. Linnakylä, P. Reinikainen, S. Sulkunen, J. Törnroos, & I. Arfmann (Eds.), *The Finnish success in PISA – and some reasons behind it.* Jyväskylä: Kirjapaino Oma OY.

Wallace, C. S., Hand, B. M., & Prain, V. (2004). *Writing and learning in the science classroom.* Dordrecht: Kluwer.

Yore, L., Bisanz, G. L., & Hand, B. M. (2003). Examining the literacy component of science literacy: 25 years of language arts and science research. *International Journal of Science Education, 25*(6), 689–725.

Chapter 4
The Effect of Embedding Multimodal Representation in Non-traditional Writing Task on Students' Learning in Electrochemistry

Murat Gunel, Sevgi Kingir, and Nurdane Aydemir

Introduction

Current movements to reform science curriculum view scientific literacy as their central goal. Many countries have made radical changes to their curricula to achieve scientific literacy for all students. Within these reform movements, curricula are designed to allow individuals to cope with the changes in science, technology and society (Hurd 1998). In Turkey, for example, there is more emphasis on scientific literacy in school science curriculum than there was before 2000. The Turkish curriculum reform movements in 2000, 2004 and 2013 concerning the elementary and secondary science curricula all emphasized scientific literacy. In the reforms of 2000 and 2004, a scientifically and technologically literate person was defined as a person who (a) understands the scientific nature of knowledge; (b) understands the basic science concepts, principles, theories and laws, and utilizes them appropriately; (c) utilizes science process skills during problem-solving and decision-making; (d) understands the interaction among science, technology, society and environment; and (e) develops his/her scientific and technical psychomotor skills (Ministry of National Education [MNE] 2000, 2004). In addition to the changes to

M. Gunel (✉)
Faculty of Education, Department of Primary Education,
TED University, Ankara, Turkey
e-mail: murat.gunel@tedu.edu.tr

S. Kingir
Faculty of Education, Department of Elementary Education, Hacettepe University,
Ankara, Turkey
e-mail: ksevgi@hacettepe.edu.tr

N. Aydemir
Alanya Faculty of Education, Department of Elementary Education, Akdeniz University,
Antalya, Turkey
e-mail: nurdaneaydemir@akdeniz.edu.tr

© Springer International Publishing Switzerland 2016   59
B. Hand et al. (eds.), *Using Multimodal Representations to Support Learning in the Science Classroom*, DOI 10.1007/978-3-319-16450-2_4

the elementary science curriculum, the high school chemistry curriculum was also revised with the same vision (MNE 2007). Although the reform movement in the Turkish curriculum aimed to promote science literacy in a large scope, important components of literacy and language appear to have received minor emphasis.

The development of scientific literacy can be achieved through the appropriate use of language because it is impossible to understand science, clearly communicate scientific ideas and make informed decisions about scientific, technical, social and environmental issues without being able to read, write, speak and write (Hand et al. 2009b). Within this framework, writing is an integral part of language and a very powerful tool for the construction of scientific knowledge. However, using writing alone is not sufficient for meaningful science learning. The quality of the writing tasks in which writing is intended to serve as a learning tool has a critical role in determining whether meaningful learning is achieved. In a meaningful learning activity, writing is a facilitator of knowledge constitution rather than taking the role of knowledge telling (Galbraith 1999). In the knowledge-telling model, ideas about a topic are simply retrieved from long-term memory and then translated into text, which, in turn, results in rote memorization (Bereiter and Scardamalia 1987). However, in a knowledge constitution process, ideas are synthesized by adding or omitting some meanings, thus the writing product is not simply a display of the knowledge stored in long-term memory (Galbraith 1999). The majority of current in-school practices utilizing writing are based on a knowledge-telling model or a more traditional writing process and there is a considerable need for approaching writing from a knowledge constitution perspective. This requires a more non-traditional implementation of writing. Non-traditional writing tasks often take different formats; such as letters, newspapers and power-point presentation to serve different purposes such as reviewing, clarifying, persuading, and for different viewers other than the teacher (Prain and Hand 1996). Students that are engaged in appropriate writing-to-learn strategies are able to think in a complex way, use language consciously and retain more knowledge (Glynn and Muth 1994; Rivard 1994). On the other hand, while there are many cases about the function and use of writing as a learning tool in the international literature, recent research conducted by Biber (2012) in Turkey, and Villalon and Mateos (2009) in Europe revealed that the majority of science teachers and high school students are not aware that writing can be used as a learning tool. Therefore, any effort to disseminate the potential learning function of the writing tasks among the learners and teachers remains to be a crucial area of investigation in science education.

Writing-to Learn Studies in Turkey

The investigation of writing-to-learn implementation can help develop strategies for extending the use of writing as a tool for learning. In his work, Biber (2012) pointed out that in the science classes in Turkey, teachers' perceived function of writing mainly focused on note taking. In the same study, teachers' lack of understanding and awareness of the pedagogical practices associated with using writing as a learning tool was attributed to their former experience with writing when they were students, the absence of relevant professional development programs and the national curriculum not placing sufficient emphasis on writing. Although recent reform movements in Turkey heavily foreground the major tenants of science literacy through changes to the curriculum and legislation, the fundamental connection between science literacy and language, particularly with regard to writing appears to have been disregarded. Therefore, in the Turkish science education literature there are calls for further research concerning the rudiments of in-school writing-to-learn implementations. On the other hand, relatively expanded international literature and majority of the Turkish literature on writing-to-learn demonstrates tendency toward certain domain in science learning such as middle school science and high school physics.

In the international literature, there are several in-depth investigations into the function of writing as a learning tool (Prain and Hand 1996). However, studies on writing in science education have only recently gained momentum in the Turkish settings and remain limited in number and scope. Findings of the studies conducted in Turkey are consistent with those of other settings. When the outcomes of traditional and non-traditional writing tasks in the Turkish educational settings were analyzed, non-traditional writing tasks were found to be more effective in supporting students' science learning compared to traditional writing tasks (Gunel et al. 2006, 2007, 2009b; Kingir 2013). However, a closer examination of research in both national and international settings indicates that they mainly focused on physics topics and on probing the effects of embedding multimodal representations in the writing tasks. These characteristics point out both the need for further studies in the areas of biology and chemistry, as well as the potential for the exploration of multimodal representations to provide a framework for an international research agenda.

Multimodal Representations and Writing to Learn

Representations in science are defined as devices used to symbolize an idea or concept in science and can take the form of text, mathematical formulas, diagrams, graphs, and simulations (Tang et al. 2014). In the language of science and science education, the term 'representation' is the subject of two synonymous but distinct research agendas, namely multiple representations and multimodal representations. The former encompasses representing the same ideas related to science through different representations as well as the relationship between students' learning of science and use of multiple representations (Prain and Waldrip 2006; Prain et al. 2009). The latter involves the integration of more than one modality to communicate these ideas. The components of various modalities such as language, depiction, formulas and symbols are either individually or simultaneously used to illuminate students' science learning (Tang et al. 2014).

As defined above, multimodal representations are modes used for the representation of scientific knowledge. Alternative modes can be defined as those that are different from the text, such as graphics, pictures, diagrams, symbols, mathematical expressions, tables or lists. However, to achieve the greatest effect, multimodal representations should be embedded in the text referring to similar conceptual ideas, rather than simply placed next to the text. Consideration must then be given to determining whether modes other than text are embedded in the text. In general, if such a mode cannot be understood without the text and vice versa, that mode is considered to be well embedded in the text. Gunel et al. (2006) suggested using multiple modes within alternative writing tasks rather than a standard format in which students summarize the unit under investigation in order to obtain a higher conceptual understanding. In another study, Gunel et al. (2009a) examined the effect of embedding multimodal representations in the writing task on the science learning of elementary students and concluded that non-traditional writing with multiple embedded modes of representation supported students' understanding of the force concept.

The use of multimodal representations in science learning, and in particular, embedding them in writing activities, is a well-established yet constantly developing strand of research. Integrating writing to learn strategies with multimodal representations provides key opportunities for students to translate between different modes in representing scientific concepts as well as between scientific and everyday language. Writing allows students to move between different modes to articulate the meaning through their own language (Hand et al. 2009a). As students move between scientific and everyday language, they re-represent the concepts using multiple modes (Gunel et al. 2006). Such a scaffolding mechanism calls for research to investigate the student choice (critical consultation) of modal representation as they try to create a coherent multimodal explanation of a scientific concept (Airey and Linder 2006).

Multimodal Representations and Writing to Learn in Chemistry Education

Recent studies in chemistry (McDermott 2009) and physics (Gunel et al. 2006, 2009b) encourage student development through writing-to-learn activities to create written products with multimodal representations embedded in the text. The results of these studies are promising in terms of increased conceptual understanding. Chemistry, by its nature, is very abstract because most of the phenomena discussed in chemistry occur at the microscopic level (Garnett et al. 1995). For this reason, presenting chemistry information by text alone is not meaningful for understanding the scientific concepts since it often results in rote memorization. Therefore, using multiple modes in representing chemistry knowledge can both mirror common practices associated with the discipline and can be a critical factor in improving student learning. For this reason, the purpose of the present study was to contribute to this developing research through the investigation of the impact of embedding multiple modes in text on student learning in electrochemistry. The present study also aimed to investigate the effect of writing-to-learn activities embedded with multiple modes of representation on 11th grade students' conceptual understanding in electrochemistry. Furthermore, guided by the current literature, the present study explored the complexity of students' writing samples when they were asked to purposefully engage with multimodal representations. Finally, the study presented the students' perspective of the process of writing and of embedding multimodal representations in text. Consequently, the study explored the following research questions:

- What is the impact of the instruction designed to encourage the use of multimodal representations in writing on students' understanding of electrochemistry topics?
- What is the impact of the instruction designed to encourage the use of multimodal representations in writing on students' writing assessment scores?
- How do treatment students perceive the value of the writing to learn activity and multimodal representation instruction?

Methods

In this study, a quasi-experimental, pre-posttest design was used. During the study, basic electrochemistry concepts of oxidation-reaction and electrochemical cells were included in the regular classroom curriculum of the chemistry course. Aligned with the above-mentioned research questions, this section provides detailed information on the participants, the procedure that was followed and the data collection process. Furthermore, data analysis is discussed in this section.

Participants

The participants of this study were a total of 57 eleventh grade students from two intact classes of a chemistry teacher attending a public high school located in Ankara, Turkey. One of the classes was randomly assigned as the treatment group and the other as the control group. There were 31 students (22 females and 9 males) in the treatment group and 26 students (13 females and 13 males) in the control group. Students' ages ranged from 16 to 17 years old. Students came from a similar socio-economic background.

Procedure

The normal instructional time for the chemistry course was three 45-min periods per week over a 4-week period. Two intact classes of an instructor who volunteered to participate in this study were assessed for the testing procedures. These classes were randomly divided into the treatment and control groups. Immediately prior to the procedures, the students were given a pretest on the electrochemistry concept (details are given in the following section) to detect any differences between the groups. Following the pretest, the intervention procedure began. The language of instruction and data collection tools used in this study was Turkish.

Students in both groups were instructed in electrochemistry concepts using the traditional instruction method that the teacher had employed in the previous years. As part of this traditional pedagogy, the teacher mainly used lecture and discussion methods, and solved problems for the students. Prior to each lesson, students were asked to read the related topic from the course book. The teacher explained each concept and asked questions to promote discussion. Towards the end of the lesson, the teacher distributed worksheets that included topic-related mathematical and conceptual questions for the students to answer. Close to the end of the 4-week period, these worksheets were scored and corrected by the teacher, answers were provided, then the students reviewed the corrections that had been made on their worksheets.

The only difference between the treatment and control groups in terms of unit activities was an additional 3-session multimodal instructional lesson provided in the treatment class. Students in the control group did not receive this instruction. While the first session was conducted outside class, the second and third sessions in relation to the writing-to-learn activity and multimodal representations were conducted during two extra class hours in order to devote equal amount of instructional time to the targeted concepts for treatment and control groups. To ensure that equal time was spent on task by both groups, the control class had extra problem solving homework and recitation session aligned with the treatment groups' time spent on writing and the multimodal representations.

During the first session of the treatment intervention, the teacher designed a warm-up activity to introduce the different modes used in scientific text to the students. For this activity, the teacher gave each student a scientific article related to recycling, which was taken from a monthly national journal on popular science targeting high school and college students. This article included different modes that might be interesting for the students and the topic of the article was not included in the chemistry curriculum. The teacher asked the students to read the article outside the class and answer the following questions:

1. What is the main idea of the article?
2. What are the strengths and weaknesses of the article?
3. Is the article easy or difficult to understand? Explain why.
4. What are the effective modes used in the article to convey the main idea?
5. If you were the author, how would you convey the main idea?

6. Who do you think is the target audience of this article? Explain your reasons for this choice.

The in-class homework introduction and brief discussion took approximately 15–20 min. It was expected that the students would spend 1–2 h a week working on the homework assignment. The second session of the treatment intervention took place within a class hour. The purpose of this session was to analyze and discuss how the common sources of scientific information embed multiple modes in text to represent this information. The researchers chose two pages from the electrochemistry chapter of a second course book written by Dursun and Kızıldağ (2005) to be used as the focus of this session for the effective use of different modes. With the guidance of the teacher, the students were asked to form their own groups and list the multimodal representations (such as mathematical formula, picture, graph and text) given in these two pages. Then, the following questions were used to lead to a whole class discussion:

1. What are the effective modes used in the written text?
2. How are the modes related to each other?
3. Which mode(s) helped us learn about the topic?
4. How should we use mode(s) together to convey the meaning properly?

In the third session, the students, as a group, were asked to develop criteria for the assessment of the effectiveness of multimodal use in a written text. They were asked to take into account what they had learned from the activities in the first and second sessions. Then, each group shared their criteria with the others. During the discussion, the teacher asked a student to write on the board the criteria that all the students and the teacher agreed on during the discussions. At the end of the session, one student copied the board work into a word processing program and sent it to the teacher and peers. The aim of these three sessions was to encourage students both to recognize effective multimodal use in scientific sources and to effectively embed alternative modes of representation in their explanatory texts.

At the conclusion of the instruction, both the control and treatment classes participated in an identical end-of-unit writing task, in which they were asked to use at least one mode other than text. The writing task was writing a letter to the 8th grade students. The students were given a handout on how to write a letter and asked to summarize the major electrochemistry concepts (such as oxidation, reduction and electrochemical cells) using multiple modes of representation. The students wrote the letters at home and submitted them to the teacher within 1 week. The students in the treatment group were also required to assess their letters based on the checklist that had been formulated as a part of the lessons on the multimodal use, make any necessary changes to their letter and submit their letters with the checklist. Following the submission of the letters, the students were administered the same unit test given at the beginning of the instruction period. Furthermore, semi-structured interviews were conducted with 12 volunteer students in the treatment group to explore their ideas about multimodal instruction and using multimodal representations in writing tasks. Since the control group did not receive the multimodal instruction, they were not interviewed.

Data Collection

Data was collected using a concept test, a writing assessment scale, and interviews with 12 students from the treatment group as explained in the following sections.

Electrochemistry Concept Test (ECT) This instrument was used to assess students' understanding of the electrochemistry concepts. ECT was developed by the researchers considering the objectives related to electrochemistry included in the national chemistry curriculum (MNE 2007) and using various sources such as course books, internet, the Student Selection Examination for admission to higher education in Turkey and related literature (e.g., Chou 2002; Sanger and Greenbowe 1997; Yılmaz et al. 2002). ECT consisted of 13 multiple-choice and 5 open-ended questions related to oxidation-reactions and electrochemical cells. Each multiple-choice item had five alternatives including one correct answer and four distracters.

Prior to the administration of ECT, two professors and two research assistants in chemistry education evaluated the test to confirm the validity of the content. In addition, two chemistry teachers and a Turkish language teacher provided an assessment and feedback on the appropriateness of the reading level of the language used in the test. Upon completion of the revisions based on these reviews, the test was administered to the participating students twice as a pre- and post-test. The coefficient of internal consistency reliability was calculated as 0.70. In the scoring of multiple choice test items, each correct response was scored as 1 and each incorrect response was scored as 0. For open-ended test items, correct responses were scored as 2, partial correct responses were scored as 1, and incorrect responses or no response were scored as 0, using a rubric describing the assessment criteria. The total maximum score of this test was 23 and the minimum was 0. ECT was administered to both groups by the teacher as a pre- and posttest during the regular class sessions, each taking 30 min. Figure 4.1 presents examples of the ECT items.

Assessment Scale for Writing The letters written by the students at the end of the instruction period were scored using the scale developed by McDermott (2009). In this scale, three sub-categories are assessed; namely text, modes and embeddedness. The first sub-category of the writing assessment is the text produced by the students. The text score was calculated as the sum of four scores obtained by assessing whether the students covered the required topics, were accurate in the scientific concepts they wrote about, developed complete descriptions and used appropriate grammar. A three-point scale was used to score each item. The score for the second sub-category of writing assessment, namely the modes, was computed by adding together the overall number of modes other than text, the number of different types of modes used, and the total number of science topics addressed using these modes. The score for the third sub-category, embeddedness, was determined by individually assessing each mode other than text and then adding these scores together to obtain the total score. The criteria for assessing each individual mode other than text were whether the mode was original, accurate, complete, next to the text that referred to it, referenced in the text, and whether it contained a caption. To ensure the inter-rater

13. If magnesium electrode decreases in mass in an electrochemical cell shown in below, which of the following is a wrong statement?

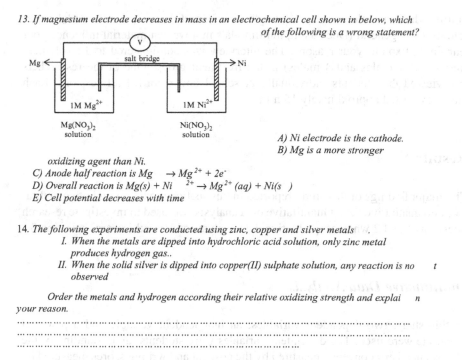

A) *Ni electrode is the cathode.*
B) *Mg is a more stronger*
        oxidizing agent than Ni.
C) *Anode half reaction is Mg* $\rightarrow Mg^{2+} + 2e^-$
D) *Overall reaction is Mg(s) + Ni* $^{2+} \rightarrow Mg^{2+}$ *(aq) + Ni(s)*
E) *Cell potential decreases with time*

14. The following experiments are conducted using zinc, copper and silver metals   .
   I. *When the metals are dipped into hydrochloric acid solution, only zinc metal produces hydrogen gas..*
   II. *When the solid silver is dipped into copper(II) sulphate solution, any reaction is no   t observed*

   Order the metals and hydrogen according their relative oxidizing strength and explai   n your reason.

....
....
....

Fig. 4.1 Examples of the ECT items

reliability of the scale used in this study, writing samples were randomly selected and independently scored by the three researchers until reaching a consensus. Following the assessment of three writing samples, the overall agreement attained by the researchers was 88 %. The remaining writing samples were scored by one researcher.

Semi-structured Interviews Interviews were used for the purpose of understanding students' ideas about the multimodal instruction and using multiple modes of representation in their writing tasks. One of the researchers constructed the interview protocol. The first draft of the protocol was evaluated both by a professor in science education to assess the appropriateness and completeness of the content in relation to its purpose and by a research assistant with a PhD in chemistry education in terms of ambiguity and leading questions. In line with the recommendations of these experts, a question was added to fit the purpose of the study and some questions were rephrased for clarity. Then, the interview protocol was pilot-tested with a student. The results of this pilot test showed no problems in terms of the clarity of the items.

The final draft of the interview protocol contained 18 questions related to the following three different activities performed by the students for the introduction of the multimodal representations: writing to learn tasks (e.g., Do you think that multimodal instruction influenced your knowledge about modes?, How?), writing a letter to the 8th grade students (e.g., Have you transferred what you learned from multimodal instruction to your writing?, How?), and the use of multimodal representations in general (Does using multimodals in a written material influence your learning?, Explain your reason). The interview was administered to 12 volunteer students (8 females and 4 males) in the treatment group. One of the researchers interviewed the students individually at school and recorded all sessions. Each interview lasted approximately 15 min.

Results

The major findings of this study reported in this section are based on both quantitative and qualitative data. Quantitative data analysis was used to investigate research questions 1 and 2 whereas qualitative data analysis addressed research question 3.

Quantitative Data Analysis

In this study, two dependent variables, an independent variable and a potential covariate were used. The dependent variables were students' understanding of the electrochemistry concepts measured by the posttest and writing scores measured by the writing assessment scale. The independent variable was the treatment condition. The potential covariate was students' prior understanding of the electrochemistry concepts measured by the pretest. A value of 0.05 was considered to be the significance level in the statistical tests. Cohen's d index was used to assess the practical significance of the results. In this index, an effect size is interpreted as *small* if $0.2 \leq d < 0.5$, *medium* if $0.5 \leq d < 0.8$, and *large* if $d \geq 0.8$ (Cohen 1988).

Pretest Analysis The mean pretest scores were found to be 4.28 in the treatment group and 3.75 in the control group. For the comparison of the mean pretest scores between the groups, one-way Analysis of Variance (ANOVA) was used. The results revealed no significant differences between the two groups in relation to the students' prior understanding of electrochemistry concepts ($F(1, 50) = 0.618, p = 0.436$).

Posttest Analysis One-way ANOVA was used for the comparison of the posttest scores of treatment and control groups to assess the impact of embedding multiple modes of representation in text on students' understanding of electrochemistry. The results showed that there was a significant difference in the mean posttest scores between the two groups in favor of the treatment group, ($F(1, 49) = 4.082, p = 0.049$, $d = 0.58$). The post-test mean score was found to be 9.79 in the treatment group while it was 8.13 in the control group.

The scores of the students in both groups were also compared in terms of the three writing sub-categories, namely text, mode, and embeddedness, using a Multivariate Analysis of Variance (MANOVA). The results showed that the effect of treatment (Wilks' Lambda$=.589$, F (3, 44)$=10.217$, $p<0.05$, $d=1.67$) was significant on the collective dependent variables. The results of a further univariate ANOVA indicated a significant mean difference between the treatment and control groups with respect to the students' text scores, $(F$ (1, 46)$=13.473$, $p<0.05$, $d=1.08$), embeddedness scores, $(F$ (1, 46)$=13.718$, $p<0.05$, $d=1.09$) and mode scores $(F$ (1, 46)$=24.488$, $p<0.05$, $d=1.46$).

Qualitative Data Analysis

Students' interview responses constituted the qualitative data in this study. Before the analysis of interview data, audiotapes were fully transcribed. Then, three randomly selected interview transcripts (25 % of all transcripts) were independently coded by two researchers. After each transcript was coded, the results were compared. Following a discussion about the discrepancies on some codes, the coders reached a consensus. When an inter-rater agreement of 92 % was achieved, one researcher continued to perform the coding of the remaining transcripts. The codes that were similar were grouped into the following three categories; activities provided within the scope of the treatment, writing task and multimodal representation (Marshall and Rossman 2006). The interview responses were interpreted based on the codes and categories as shown in Table 4.1.

Activities Provided Within the Treatment The students were asked about the activities provided as part of the treatment (receiving information about the multimodal representations, understanding the use of multimodal representations within the text through a group discussion and the formulation of a checklist). All the students that were interviewed agreed that the modes other than text did not capture their attention before these activities. One fourth of the students claimed that they were using some of the modes without being consciously aware. All the interviewed students stated that they became more alert to the multimodal representations within the text as a result of the lesson on multimodal representations. Many students (75 %) reported that the group discussion and the formulation of a checklist were beneficial to their understanding of multimodal representations.

Writing Task All the interviewed students stated that writing a letter contributed to their learning. Students also commented that they recognized the inadequacies in their understanding when they were writing, and they closed the gaps in their knowledge through writing. Some (25 %) suggested that, learning a concept is not possible without internalizing it. One student explained her view as, "I used to confuse the terms; oxidation and reduction. When I was writing, I thought about the best way to teach these concepts. Then, I found an analogy and tried to teach them using that analogy. Actually, it was very meaningful…I also learnt. First, I learnt by myself, then I taught (other students)". Students' attitudes towards the writing activity were generally positive. Half of the interviewed students stated that writing a letter to younger peers was entertaining, interesting and effective in reinforcing

Table 4.1 Categories and codes with examples

Category	Code	Example
Evaluation of activities provided within the treatment	Recognition	"Before this instruction I did not look at the modes when I was reading a text."
	Useful	"As a result of this instruction, I now know that modes can be helpful in my learning."
Ideas about the writing task	Contribution to learning	"…I learnt more because I thought more when I was writing."
	Attitude	"The writing activity was interesting, and entertaining."
	Knowledge transformation	"…I tried to use less scientific terms. I explained the concepts and added annotations…"
	Self-reflection	"While I was assessing my own letter using our checklist, I recognized that the captions of the modes were missing. I added the captions myself and then explained the modes in the text in detailed"
Perceptions on multimodal representation	Relationship between multimodal representations and learning	"The more you use modes in your writing, the more you learn about the concepts…. because text becomes boring after a while."
	Awareness	"I did not use the modes when I was taking notes but now I am trying to use them."
	Knowledge retention	"…The learning material including various modes is very helpful in learning because it helps retain knowledge."

their learning. Two students thought themselves as a teacher when writing the letter. Furthermore, all the students stated that they attempted to use a language appropriate for the eighth graders to help them better understand the concepts. The interviewed students indicated that they tried to avoid using too many scientific terms in their writing; instead, they used everyday language to explain the concepts. A small number of students (17 %) used the term 'prior knowledge' in their responses. For example, one student stated, "I thought of myself as an 8th grader, as a student not knowing these concepts…They could lack some knowledge related to these concepts. I took into consideration the things they may not know, that is, I reminded the reader about the previous knowledge…" The students also found the self-assessment of their letters very helpful because through this activity they became aware of what was missing in the letter and tried to write it properly. Some made changes to their letters while others did not. The students who did not make any changes claimed that no conceptual information was missing since they paid attention to the checklist when writing the letters. Some students (25 %) stated that self-assessment increased the objectivity of the written material. For example, a student expressed her view as, "…I explained some of the mathematical expressions in more detail… I knew what I meant but if I did not explain it clearly to the audience, I could not guarantee that they would understand it properly…"

Multimodal Representation The analysis of the writing samples and interviews provided some evidence that students learnt how to use alternative modal representation since they used them to articulate scientific information. For example, some students used graphics, pictures and diagrams interchangeably in addition to proper textual explanations. The analysis of the interview responses indicated that almost all the students considered the use of pictures, diagrams, graphics, symbols and mathematical expressions to be helpful in learning the electrochemistry concepts, which was the objective of the writing assignment. In addition, students preferred using multiple modes within text to using text only. For example, students stated, "alternative modes are visual...easily remembered", "alternative modes facilitate learning", and "alternative modes attract my attention, and further enhance my learning". A student stated that using alternative modes was central to learning chemistry concepts. Most students (83 %) commented that after participating in this activity they began to pay closer attention to the use of multiple modes of representation in other sources when studying on their own. A few students (17 %) stated that before this study, they had just begun to use multimodal representations. Many students (75 %) indicated that they used lists, tables, pictures, graphics and diagrams in their notes when studying on their own.

Discussion

This study investigated the effect of using multiple modes of representations within text on the conceptual understanding of electrochemistry of the 11th grade students. For the investigation of the first research question, namely *"What is the impact of the multimodal representation instruction and embedding multimodal representations in writing on students' learning of electrochemistry topics?"* a quasi-experimental approach was adopted using the pre and post-test design. Upon completion of the intervention, the understanding of electrochemistry of the treatment and control groups was evaluated through an instrument containing multiple-choice conceptual questions. It is important to note that both groups were asked to complete a similar end-of-unit writing to learn activity, writing a letter to the eighth graders explaining what they learned from the electrochemistry lesson. However, the treatment group had a series of instructional activities emphasizing the meaning, function and use of modal representations in communicating science whereas the control group spent the same amount of time on problem solving about electrochemistry.

The results indicated that there was no significant difference in pretest scores between the treatment and control groups. That is, at the beginning of the intervention the students in both groups were similar with regard to their knowledge of electrochemistry. On the other hand, significant differences were found in the post-test mean scores between the groups, in favor of the treatment group. The size of the mean difference was 0.58, indicating a medium effect size according to Cohen's criteria (Cohen 1988).

The significant difference in the learning outcome of the treatment and control groups supports the idea that when students' understanding, awareness, and ability to effectively use multimodal representations are scaffolded, their conceptual understanding of the targeted concepts in the unit studied improves. Enhancement in the students' understanding of chemistry indicates that the findings are consistent with those obtained from the previous studies on multimodal representation conducted mainly in the physics area focusing particularly on mechanics and electricity (Gunel et al. 2006, 2009b; Hand et al. 2009a). The findings presented here are important not only due to the limited amount of research on multimodal representation with regard to chemistry topics but also considering its potential value in facilitating the learning of abstract concepts in chemistry at the microscopic level (Garnett et al. 1995). The post-test comparison results of this study demonstrated that in the treatment group, students' understanding of electrochemistry increased at least partly due to the use of multimodal representations. Encouraging students to understand and translate modal representations in chemistry and using them to communicate scaffolded conceptual understanding. From a pedagogical perspective, implementing interventions as used in this research have potential benefits and practical considerations for teachers.

Introducing modal representations with popular articles, assigning homework such as writing tasks, and holding in-class discussions about the function and use of the modal representations were found to have a positive impact on the learning process and the perceptions of students. Teachers are constantly in search for improved pedagogical implementations. The multimodal representation instruction outlined in this study can be an appropriate way for teachers to widen their pedagogical repertoire since it is not only well perceived by students but also enhances the students' understandings about the concepts in chemistry.

The second research question, namely *"What is the impact of multimodal representation instruction and embedding multimodal representations into writings on students' writing assessment scores"*, was investigated analyzing the text of the letters written by the students in both groups. The results indicated significant differences in the end-of-unit writing scores between the two groups, in favor of the treatment group. The size of the observed mean differences between the groups in terms of text scores, embeddedness scores and total mode scores were 1.08, 1.09 and 1.46, respectively, indicating a large effect size according to Cohen's criteria (Cohen 1988). It is clear that the treatment intervention of multimodal representation instruction played a role in effect size being large. However, it is important to note that short and brief interventions related to multimodal representations not only led to a large effect in mode scores but also in embeddedness scores. That is, increasing the awareness and understanding of modal representations resulted in generating more complex and extended writing. Such an impact indicates not only the enhancement in science literacy skills as discussed by Glynn and Muth (1994) and Rivard (1994) but also the opportunity to scaffold conceptual understanding through writing as argued by Prain and Hand (1996). Putting together our findings and the findings of the previous research, we can conclude that implicit instruction in multimodal representation creates more ground for effective writing-to-learn implementation in science classes.

The third research question, *"How do treatment students perceive the value of writing to learn activity and multimodal representation instruction?"*, was investigated to seek a deeper understanding about the impact of multimodal instruction on learning from the students' perspective. The results of the interview revealed that students were indeed familiarized with the multiple modes as a result of the instruction. In their responses to the interview items, students described their own learning as the development of their thinking. Such findings not only point out the value of writing-to learn activities at a metacognitive level as discussed by Gunel et al. (2006) but also add a new dimension and complexity to maximize learning outcomes with embedded multimodal representations in non-traditional writing. That is, at a metacognitive level, students indicated that they were more aware of their own learning, and they articulated their thoughts and understandings when they used modal representations in their letters to the younger students. Such an awareness and understanding also have a close relationship with Airey and Linder's (2006) call for critical consultation of modal representations. In their study, when the students were asked to evaluate written materials to construct meaning using modal representation, they became more critical about how to construct meaning through mediums such as text, graphical and mathematical representations. Thus, as they generated their text through their own understanding, they became more aware of their ability and conciseness about the critical consultation of modal representations. Similarly, in the current research, the findings of the interviews indicate such self-conciseness about critical consultation; however, there is a need for further studies to investigate the dimension of consultation of modal representation.

Conclusion

The results of this study support the idea that using multiple modes of representation within text in a non-traditional writing task results in a higher conceptual understanding, improved science literacy skills, positive attitudes toward science and enhanced awareness about communicating scientific ideas. Some of these findings, particularly those related with the conceptual understanding of science are consistent with the findings of the previous research (Gunel et al. 2006, 2009b; McDermott 2009). Although non-traditional writing and modal representations have benefits in science education and science literacy, they have limited implementation in educational settings. From a Turkish educational perspective, using non-traditional writing tasks are not common for several reasons including the lack of understanding, information and guidance about how to effectively implement them. Contrary to traditional writing, the teachers and students are not familiar with the purposes and potential benefit of alternative writing tasks. Similar to the findings and suggestions of Biber (2012), the current research advocates a shift from traditional writing tasks to non-traditional writing tasks supported with curricular revisions. Furthermore, since some teachers are not experienced in implementing appropriate writing-to-learn strategies and promoting multimodal representations in the classrooms, pre-service and in-service teacher training can facilitate this shift focusing on the development of teaches' pedagogy, beliefs and attitudes about the writing-to-learn strategies with embedded multimodal representations. Finally, it is crucial to disseminate the information about the associated benefits such as skills development and science learning outcomes in high-stake test-based educational settings, such as the system in Turkey, since academic test performance is the driving sense for in-class implementations and pedagogical approaches in such systems. Therefore, teachers have a tendency to focus on drill and practice approach placing too much emphasis on problem solving. Broadening the teachers' pedagogical repertoire with non-traditional writings and multimodal representations may not only improve test performance but also scaffold science literacy skills of the students.

References

Airey, J., & Linder, C. (2006). *Languages, modality and disciplinary knowledge.* Paper presented at the ICLHE 2006: Integrating Content and Language in Higher Education, University of Maastricht, Maastricht, Netherlands.

Bereiter, C., & Scardamalia, M. (1987). *The psychology of written composition.* Hillsdale: Lawrence Erlbaum.

Biber, B. (2012). *Fen ve teknoloji öğretmenlerinin yazmaya dair algıları ve öğrenme amaçlı yazma aktivitelerini uygulama düzeyleri* [Science teachers' perceptions toward writing and levels of writing-to-learn implementations]. Unpublished master thesis. Atatürk Üniversitesi, Erzurum, Turkey.

Chou, C. Y. (2002). Science teachers' understanding of concepts in chemistry. *Proceedings of the National Science Council, Republic of China (Part D), 12*(2), 73–78.

Cohen, J. (1988). *Statistical power analysis for the behavioral sciences* (2nd ed.). Hillsdale: Lawrence Erlbaum Associates.

Dursun, M. F., & Kızıldağ, G. (2005). *MEB lise kimya 2 ders kitabı* [MNE tenth grade chemistry textbook]. Ankara: Saray Matbaacılık.

Galbraith, D. (1999). Writing as a knowledge-constituting process. In D. Galbraith & M. Torrance (Eds.), *Knowing what to write: Conceptual processes in text production. Studies in writing* (Vol. 4, pp. 139–160). Amsterdam: Amsterdam University Press.

Garnett, P. J., Garnett, P. J., & Hackling, M. W. (1995). Students' alternative conceptions in chemistry: A review of research and implications for teaching and learning. *Studies in Science Education, 25*, 69–96.

Glynn, S. M., & Muth, K. D. (1994). Reading and writing to learn science: Achieving scientific literacy. *Journal of Research in Science Teaching, 31*(9), 1057–1073.

Gunel, M., Hand, B., & Gunduz, S. (2006). Comparing student understanding of quantum physics when embedding multimodal representations into two different writing formats: Presentation format vs. summary report format. *Science Education, 90*(6), 1092–1112.

Gunel, M., Hand, B., & Prain, V. (2007). Writing for learning in science: A secondary analysis of six studies. *International Journal of Science and Mathematics Education, 5*(4), 615–637.

Gunel, M., Atila, M. E., & Buyukkasap, E. (2009a). The impact of using multi modal representations within writing to learn activities on learning electricity unit at 6th grade. *Elementary Education Online, 8*(1), 183–199.

Gunel, M., Uzoglu, M., & Buyukkasap, E. (2009b). Effects of using writing to learn activities on learning force unit in the primary education level. *Gazi University Journal of Gazi Educational Faculty, 29*(1), 379–399.

Hand, B., Gunel, M., & Ulu, C. (2009a). Sequencing embedded multi-modal representations in a writing to learn approach to the teaching of electricity. *Journal of Research in Science Teaching, 46*(3), 225–247.

Hand, B., Norton-Meier, L., Staker, J., & Bintz, J. (2009b). *Negotiating science: The critical role of argument in student inquiry, grades 5–10*. Portsmouth: Heinemann.

Hurd, P. D. (1998). Scientific literacy: New minds for a changing world. *Science Education, 82*, 407–416.

Kingir, S. (2013). Using non-traditional writing as a tool in learning chemistry. *Eurasia Journal of Mathematics, Science & Technology Education, 9*(2), 101–114.

Marshall, C., & Rossman, G. B. (2006). *Designing qualitative research* (4th ed.). Thousand Oaks: Sage.

McDermott, M. A. (2009). *The impact of embedding multiple modes of representation on student construction of chemistry knowledge*. Unpublished doctoral dissertation. University of Iowa, Iowa City, Iowa.

Ministry of National Education. (2000). *National education at the beginning of 2001*. Ankara: Ministry of National Education.

Ministry of National Education. (2004). *Elementary science and technology curriculum*. Ankara: Ministry of National Education.

Ministry of National Education. (2007). *11th grade high school chemistry curriculum*. Ankara: Ministry of National Education.

Prain, V., & Hand, B. (1996). Writing and learning in secondary science: Rethinking practices. *Teacher and Teacher Education, 12*, 609–626.

Prain, V., & Waldrip, B. (2006). An exploratory study of teachers' and students' use of multi-modal representations of concepts in primary science. *International Journal of Science Education, 15*, 1843–1866.

Prain, V., Tytler, R., & Peterson, S. (2009). Multiple representation in learning about evaporation. *International Journal of Science Education, 31*(6), 787–808.

Rivard, L. P. (1994). A review of writing to learn in science: Implications for practice and research. *Journal of Research in Science Teaching, 31*(9), 969–983.

Sanger, M. J., & Greenbowe, T. J. (1997). Common student misconceptions in electrochemistry: Galvanic, electrolytic, and concentration cells. *Journal of Research in Science Teaching, 3*(4), 377–398.

Tang, K. S., Delgado, C., & Moje, E. B. (2014). An integrative framework for the analysis of multiple and multimodal representations for meaning-making in science education. *Science Education, 98*(2), 305–326.

Villalon, R., & Mateos, M. (2009). Secondary and university students' conceptions about academic writing. *Infancia y Aprendizaje, 32*(2), 219–232.

Yılmaz, A., Erdem, E., & Morgil, İ. (2002). Students' misconceptions concerning electrochemistry. *Hacettepe University Journal of Education, 23*, 234–242.

Chapter 5
Using Multimodal Representations to Develop Scientific Literacy in South African Classrooms

Mary Grace F. Villanueva

Over the past 3 years, the Department of Basic Education (DBE) in South Africa has released the National Diagnostic Report of Learner Performance on the National Senior Certificate (NSC) examinations. The report presents a comparative analysis of learner results since 2010, trends in learner enrolment, a general overview of performance per subject area and analysis of each examination question. Findings are used to identify challenges and offer recommendations to improve pedagogy, educational policies and professional development opportunities for teachers. With respect to science education, in 2013, 51 % of the grade 12 population wrote the Life Sciences examination and 31 % wrote the Physical Sciences examination. And while the South African DBE celebrated an increase of 'passing[1]' results in 2013, closer inspection of the statistics shows that only 28 % and 26 % of learners who wrote the Life Sciences and Physical Sciences examinations (respectively) received scores of over 50 %. The diagnostic report stated that the primary challenge to learners' understanding was their failure to know scientific vocabulary and definition of concepts. Secondary problems noted included failure to comprehend and analyse the problem [within the question], poor mathematical and graph reading skills, and the inability to express themselves clearly in the short answer questions. In line with ongoing research on multimodal representations, researchers such as Klein and Kirkpatrick (2010) contend that thinking and working with various representations including text, mathematical equations, and graphs are integral for reasoning and understanding resolved concepts in science. In other words, the challenges that the grade 12 learners faced in reading the graphs, understanding the problem to solve, or communicating their ideas appear to be critical aspects related

---

[1] Over the past 10 years, passing scores in South Africa have dropped to 50 %, then 40 % and now 30 %. According to the Department of Basic Education, the lower pass rate provides access to further education and opportunities to a greater number of learners.

M.G.F. Villanueva (✉)
University of Witwatersrand, Johannesburg, South Africa
e-mail: Villanueva.MG@dbe.gov.za

© Springer International Publishing Switzerland 2016   77
B. Hand et al. (eds.), *Using Multimodal Representations to Support Learning in the Science Classroom*, DOI 10.1007/978-3-319-16450-2_5

to their failure to 'know' scientific vocabulary and concepts. The challenges cited in the report, in addition to the dismal number of learners pursuing or becoming proficient in the sciences, appear to be persistent problems which have negative consequences for improving scientific literacy in South Africa. At the same time, these challenges highlight the need to find new ways to improve the scientific literacy of South African learners.

Improving Scientific Literacy

The notion of scientific literacy in South Africa has emerged largely due to the government's acknowledgement of the role that science and technology play in economic growth, employment creation, social redress and social development (Department of Arts, Culture, Science and Technology 1996). While natural resources and agriculture have traditionally been pillars of the country's economy, the Department of Science and Technology's Ten Year Plan for South Africa (2008–2018) outlines the shift from a resource-based economy towards the development of a knowledge-based economy that "must help solve society's deep and pressing socioeconomic challenges" (Department of Science and Technology 2007, p. 1). Explicit in the plan is the increased development of human capital in higher education and careers in science and technology. Yet, one of the greatest challenges to the plan is the fact that South Africa currently has a deficit of qualified and skilled people in science and technology to consolidate such a knowledge-based economy (Reddy et al. 2009).

Over the past 20 years, the national government has attempted to address educational disparities through systematic curriculum reform and resource provision. After 1994, issues of access and participation in education were addressed by developing a unified system, establishing a National Qualifications Framework, and introducing a national curriculum that emphasized the development of home and English language skills, as well as recognition of the diversity of cultures and worldviews. The transformative efforts also reflect the nations' economic and human development strategy, which emphasizes the centrality of science and mathematics and recognizes that the development of mathematical, scientific and technological skills require intervention at the school level (Reddy 2006). However, despite the government's reconstructive policies and efforts to improve science and mathematics education, the majority of South African schools still face crippling backlogs of resources, infrastructure, and qualified teachers, all of which are necessary conditions to improve participation and achievement in science and mathematics (Reddy 2006, p. 76).

There is a dire need for an improvement in science and mathematics, as shown by the poor academic results achieved in schools, particularly in science and mathematics (Christie et al. 2007; Fleisch 2008; Taylor and Vinjevold 1999). Research into matriculant pass rates illustrate that between 1999 and 2004, an average of 4.4 % of grade 12 learners scored well enough in mathematics to gain entry into natural sciences programmes at the university level (Kallaway 2007) and, over 10 years later, these figures have not changed. The findings of the Third International and Mathematics and Science Study in 1998, and the Trends in Mathematics and Science Study in 2003 (both referred to as TIMSS), revealed that of the 50 participating countries, South African grade 8 learners were the lowest scoring performers in almost all test items in mathematics and science, well below international benchmarks. The Progress in International Reading Literacy Study (PIRLS) 2006 indicated that South African learners in Grades 4 and 5 did not have the literacy competencies required for the successful transition to reading-to-learn in the Intermediate Phase (Zimmerman et al. 2009).

Science and Language

In South Africa the most effective route for literacy and language development, specifically for second-language English learners, continues to be a contested issue. In a recent report, Heugh (2013) contends that the emphasis of learning whole words and phrases in meaningful contextual situations, otherwise known as the 'whole language' approach, has been ineffective for teaching in countries where English is not the dominant language in learners' academic or home environments. Heugh (2013) suggests that children who come from homes with limited literacy practices would be served best by teaching reading that corresponds to the sounds and symbols from the alphabetic writing system. In contrast, other South African researchers have demonstrated that whole language and varied discursive practices in science and mathematics had a positive impact on learners' reading, speaking and listening skills (Webb and Mayaba 2010), problem solving abilities (Villanueva 2010; Webb and Webb 2008), and writing skills (Villanueva 2010). While multimodal use was not the specific focus of these studies, the researchers suggested that the increased use of textual and visual representations by teachers and learners was critical in the process of meaning-making in the classroom.

Aims and Limitations of This Chapter

In order to understand the possible impacts on student learning of the use of multiple and multimodal representations in science, this chapter attempts to provide context for South African science and language practices, review how the use of representations have been researched in schools and universities, and highlight two studies relating to uncovering the mental and linguistic representations of grade 6 ELL learners in the Eastern Cape of the country. As such, this chapter is not meant to serve as a definitive representation of multiple and multimodal representation use in science education in South Africa, rather the chapter should be viewed as a platform to review 'what we know' and to engage in critical discussions regarding 'what we still need to find out'.

In light of this, the intent of this chapter is to focus on the use of representations in science education in South Africa. Critical to understanding how the use of multiple and multimodal representations are used to build science and language development in South Africa, the review of research and discussions related to the studies presented in this chapter are guided by the following questions:

1. What is currently understood by the South African science education research community about multiple and multimodal representations?
2. What are significant areas of research and research findings related to the multimodal representations use that may be unique to the social and cultural context of South Africa?
3. How do these theoretical perspectives and research findings lend to improving science instruction for South African learners?

In order to address these questions, this chapter is structured in three sections. The first section is a review of science education research related to the use of representations in South African schools and universities and focused on the scholarly work published within the *African Journal of Research in Science, Mathematics and Technology Education*. The second section is framed within the context of current theoretical and practical perspectives in language and science education. Given these understandings, two studies which focus on understanding the mental and linguistic representations that grade 6 ELL learners use in textual modalities are presented as a way to further explore the intersection of theoretical ideas in practical classroom settings. The concluding section discusses the implications the ideas proposed in this article have on research and pedagogical practices in science education.

Literature Review

The following review of literature attempts to summarize what research has been conducted in the field of multiple and multimodal representations in science education in South Africa, as well as investigate common themes of study and findings in the field of representations in science education that may be unique to the social and cultural context of South Africa. The scope of this review is limited to research conducted in South African schools and universities during the last 14 years (2000–2014).[2] The review includes peer-reviewed journal articles that fit the given criteria and focuses on studies relating to teaching and learning science for learners across all levels, including pre-service and in-service teachers. The identification of literature was first conducted through an education database search (e.g. ERIC, EBSCOHOST), as well as general search engines such as Google scholar. Initially

---

[2] The early part of the millennium was a critical period for South African education, particularly because the country was in early stages of democracy and the curriculum was designed to be reflective of the goals of the new constitution.

key words such as multimodal, representations, science education, Africa and South Africa were used; however, the search yielded limited results and studies that were not appropriate for the scope of the review. The focus was then narrowed to the leading African journal for STEM education, *African Journal for Research in Science, Mathematics and Technology Education*. Using the key words and journal repository, the search once again yielded limited results. The next approach was to review the titles of each article manually, honing in on studies related to science education and the use of multiple and multimodal representations. Mathematics and technology education studies were omitted from the search unless mathematics and technology were used to supplement the teaching and learning of science.

The search yielded 28 articles which were categorized based on the following characteristics: (a) Modality or the types of modality which was the focus of the study (e.g., mental, linguistic, visual, symbolic, physical); (b) level of participants (e.g., primary school, high school, tertiary, pre-service teachers, in-service teachers); (c) science specialization (e.g., physics, chemistry, biology, earth science, 'natural science' (primary school science)); and (d) what type of research was conducted (e.g., descriptive-qualitative, descriptive-quantitative, correlational/regression analysis, quasi-experimental, experimental, meta-analysis)

Perspectives in Science Education Literature in South Africa

Although there were no studies in AJRMSTE between 2000 and 2014 which explicitly focused on furthering the field of multiple representations or multimodal representation competency research, there were, however, a number of studies that examined the mental, linguistic, visual, symbolic and physical representations used in teaching and learning. Of the 28 articles reviewed, 12 of the studies examined one specific type of representation, namely the mental representations that learners and teacher have of a particular scientific concept. Subcategories used to classify mental representations included 'conceptions' (Lemmer 2011; Mammino 2001; Mbano 2004) and 'knowledge' such as, indigenous (Govender 2011), cultural (Dube and Lubben 2011) and everyday knowledge (Stears et al. 2003). Cognitive processes such as argumentation (Ramorogo and Ogunnyi 2010; Qhobela and Moru 2011) and problem solving (Albers et al. 2008; Alant 2004; Gaigher et al. 2006) were studied as means for investigating teachers' and learners' mental representations.

Seven studies under the category of mental representations used two (n=5) or more (n=2) modalities to integrate the components of the representations to assist in meaning-making. The most common form of multimodal use incorporated the mental and linguistic aspects of representation. Mayaba et al. (2013), as well as Villanueva and Webb (2008), investigated the role of writing as a knowledge constituting process for grade 6 learners. In 2010, Webb and Mayaba reported on a scientific literacy model (Villanueva 2010) that incorporated the multimodal processes of reading, writing, talking and 'doing' in Eastern Cape schools. Writing was also used in conjunction with concept cartoons, and argumentation for Webb et al. (2008) to further grade 9 learners' ideas about energy. Green's (2007) study which explored learner discourse and science learning in the context of microcomputer-based laboratory (MLB) collaborative learning activities suggested that the MLB environment was useful for engaging learners in discursive interactions which, in turn, positively impacted learning.

The research on visual representations often encompassed physical representations such as Braund et al. (2013) work with student teachers and their use of drama to teach science concepts with grade 6 and 7 learners. The researchers' reported that the four drama lessons demonstrated a positive impact of organization of simulated drama from the student teachers, but the student teachers often failed to link scientific phenomena, concepts and processes with learners' simulated actions. The researchers state that the "student teachers often over directed the drama and restricted greater learner autonomy and space for open dialogic classroom discourse that might have helped embed knowledge." (p. 9)

In a study related to informal learning spaces, Lelliot (2010) examined grade 7 and 8 learners' concepts of spatial scale before, during and after a visit to an astronomy centre. The participants in the study constructed personal meaning maps (a variant of concept maps generally used in informal learning institutions; see Adelman et al. 2000; Falk 2003), related to their prior knowledge of astronomy. The maps, coupled with semi-structured interviews, were used as research instruments to assess a "composite estimation of a student's knowledge of the size of the Sun, stars, the Moon and other heavenly bodies as well as his or her understanding of distance within the solar system and beyond" (p. 26). Within this study, multimodal representations were steeped in the research methodology, as well as pedagogical tools used to teach students about spatial scale. As research instruments, visual and textual representations (e.g., personal meaning maps) and linguistic representations (e.g., verbal responses in the interviews) were used in conjunction to unpack the participants' knowledge and understanding of the scale. As pedagogical tools, the participants engaged in several visual representations: a live audiovisual star show (star projections on the dome of the building) narrated by an astronomer; a scale model of the solar system whereby the participants served as the sun and the first six planets and were spaced accordingly; live projection of sunspots; and moon models. Using the personal meaning models and interview responses, the participants were scored (1-low to 3-high) on (1) their knowledge of spatial scale and (2) knowledge of relative sizes of sun and moon. In both criteria, there was an increase of level 3 understanding after the visit to the astronomy centre. Lelliot (2010) attributed students' improved understanding to the variety of experiences and suggests that linking different aids in "helping students to restructure their own knowledge and partake in meaningful learning." (p. 32)

The final study in the area of visual representations was Dempster and Stears' (2013) work on using drawings to investigate students' understanding of their internal anatomy. Their sample of grade 9 students was representative of boys and girls of middle and lower socioeconomic contexts. The study produced two interesting findings: (1) that boys were more likely to represent systems and organs visually than girls, and (2) there were no significant distinctions in the drawings of learners from middle-class schools and learners from low socio-economic status schools. The authors highlighted the direct relationship between socioeconomic status and the use of English in those communities, and outlined the implications of using traditional, text-based assessments in English in multilingual environments.

Two studies examined the use of symbolic representations in algebraic sign conventions (Govender 2007) and data representation (Wessels et al. 2006). Govender's (2007) investigation of physics pre-service teachers understanding of algebraic sign conventions in vector-kinematics identified five conceptions and "hierarchal relations". In Wessels et al. (2006) research, the authors claimed that grade 4–7 students' difficulty with the trans-numeration of numerical data resulted in the use of inappropriate arrangement strategies in the numerical task.

Finally, in the area of physical representations, Meiring and Webb (2012) used toys as an educative curriculum material with pre-service education students. Using a pre-post design, the authors found that the use of toys improved learners conceptual knowledge and confidence to teach the scientific concepts associated with the toys. Their statistical analysis demonstrated a positive correlation ($r = 0.69$) between the subject content knowledge and confidence data.

Discussion

The review of these studies demonstrates the interest in and the need to understand how the diversity of South African learners conceptualize and communicate scientific ideas. What appears to be missing in the literature, however, is how these representations can be used in the classroom along with other modal representations to produce meaning. In other words, it may be productive to investigate timescale (Lemke 2000) and compositional grain size (Tang and Moje 2010) in South African classrooms. Tang et al. (2014) suggest that the analyses of shorter temporal scales of a few seconds or a few minutes are needed to make sense of elements within a representation. When investigating these elements, greater attention should be made to the use multimodal representations to help bridge students' current representations and to develop their scientific ideas, as well as their English language proficiency. Additional research is required on how South African learners and teachers integrate and embed modalities such as language (home and additional language), mathematical symbolism (equations, graphs), and graphical representations (photographs, drawings).

Understanding Representations in South Africa

In the epilogue of the *Research in Science Education* special issue of multimodal representation research, Yore and Hand (2010) outlined the critical relationship between multiple representations, multiple modality, and scientific literacy for all.

The authors suggested that Klein's (2006) second-generation cognitive position may provide a richer alignment between the cognitive actions required to build multimodal representation competence and current perspectives of learning advocated by the learning sciences (Yore and Hand 2010). The authors posed questions such as, "What knowledge bases are needed?" and "Whose knowledge is being engaged?", and while these questions are critical for understanding the dynamics of science learning as a whole, these questions are particularly important for many post-colonial and developing countries, where indigenous knowledge systems conflicts with modern western science or where the home language of the majority of the learner population differs from the established language. According to South African researchers, learners' fluency in English – in other words, how their ideas in their home language are represented in a second language – is the most significant factor in learning science and mathematics (Howie and Plomp 2005).

Representing Home Language

The current language in education policy in South Africa allows schools to select their own language of learning and teaching and, as an extension to this policy, requires schools to address the principle of additive bilingualism which involves the maintenance of home language and access to an additional language. Over the past 15 years, more instructional tools such as textbooks, reading books, and posters have been developed and utilized to meet the language needs of speakers of 11 of the official national languages; however, the majority of teaching and learning materials used in South African schools are printed in English or Afrikaans. South African researchers have viewed the practice of engaging in English or Afrikaans text particularly problematic as these languages are not the primary languages of the majority of learners and teachers (Setati et al. 2002).

In addition to the challenges of teaching and learning in a different language, learners in South Africa are also forced to grapple with utilising different *types* of languages each day. The casual and informal verbal communication at home or within personal social circles are unlike the instructional language that happens between teachers and peers in the classroom. Furthermore, the language of science is yet another context in which learners are challenged to use specialised language to communicate various content and process skills (England et al. 2007). In the course of moving from informal to instructional or scientific language, learners are continually engaged in language 'border crossing' (Yore and Treagust 2006). The three-language problem and border crossing exists for most science language learners, but the problem is often magnified for learners who are taught in a language, which is not their mother tongue. For example, in the Eastern Cape, isiXhosa is the widely spoken indigenous language and home language to 83.8 % of the population, yet the official medium of instruction in the majority of schools from the beginning of grade 4 (ages 9–10) to grade 12 is English or Afrikaans (Probyn 2004). Learners who have very little or no exposure to English or Afrikaans are placed at a serious disadvantage when one considers that these learners have minimal opportunity to speak, read or write in a second language. Furthermore, researchers (Zimmerman et al. 2009) stress the relationship between under-achievement and having being taught and assessed in a second or additional language.

In efforts to address issues of language and learner performance, the practice of code switching is often used in South African classrooms. The principle of the practice is to expose learners to words, concepts and ideas in English or the language of instruction, yet reinforce these ideas with the learners' language of familiarity (Setati et al. 2002). This practice allows learners to communicate freely and places less emotional burden if learners are not able to pronounce English words (Probyn 2001). However, while code switching is meant to assist in strengthening learners' language abilities and cognitive understanding, critics warn that this practice provides little opportunity for learners to build a firm academic or cognitive foundation in their mother tongue or the additional language (Heugh 2000; Cummings 1984). Furthermore, a greater instructional and time-consuming burden is placed on teachers to translate materials and concepts from the official language of instruction to the language learners understand (Holmarscottir 2003). Notwithstanding the criticisms of code switching, the practice is regarded as an important aspect of a bilingual or multilingual classroom. The practice is especially valuable, yet challenging in science classrooms in which teachers have the responsibility of moving learners from informal spoken language to formal written language and discourse specific talk (Gee 2002).

Improving Scientific Literacy

In an attempt to unpack "What knowledge bases are needed?" and "Whose knowledge is being engaged?" within a South African context, this section presents a small empirical study that is underpinned by learner-constructed representations and focuses on the analysis of learners' textual work. The study was conducted with grade 6 teachers and learners in one urban and one rural setting in the Eastern Cape, South Africa. Of the nine South African provinces, the Eastern Cape is characterized by having the second largest learner population and the greatest number of schools.[3] In addition, the province is the poorest in the country and continually ranks the lowest in learner achievement (Statistics South Africa 2011). In order to understand the teaching and learning challenges and ways in which to improve science education in the Eastern Cape, the study presented in this section investigated grade 6 science teachers' use of a scientific literacy model whereby the teachers engaged their learners' in the discourse practices of reading, writing, 'doing', and engaging in argumentation and whether these practices would impact and potentially improve science learning in the classroom. The activities promoted in the model serve to demonstrate how learners' thinking and language abilities can be

---

[3] Eastern Cape: 1, 910,265 learners; 5589 schools.

established through various representations, as well as serve as a starting point to consider the way multimodal representations can be used or designed to strengthen teaching and learning practices in science classrooms.

In this model, specific activities are related to each component of the strategy. For example, in the line of learning, teachers may bridge the ideas and words generated by learners during investigations to the scientific community's accepted views and vocabulary of the target concept. The new knowledge that is gained is applied to new contexts and may prompt new questions. Activities in the line of learning may be achieved through teacher-lead discussions, demonstrations and additional reading.

The underlying assumption in Fig. 5.1 is the idea that learners are engaged in "doing" or constructing their understanding through reading, writing, talking, and engaging in inquiry and argumentation. Teachers are able to identify "gaps" or misconceptions in learners' understanding, thus re-evaluating their pedagogical approaches. Learners, on the other hand, can identify areas which require additional attention and where they should revise their understanding. Finally, through formalised communication, such as writing reports or presentations to a specified audience, teachers can summatively assess whether learners were able to meet the proposed outcomes of the lesson or learning strand.

Participants and Setting The study took place in one rural and one township area in the Eastern Cape. The schools in each setting were broadly matched as under

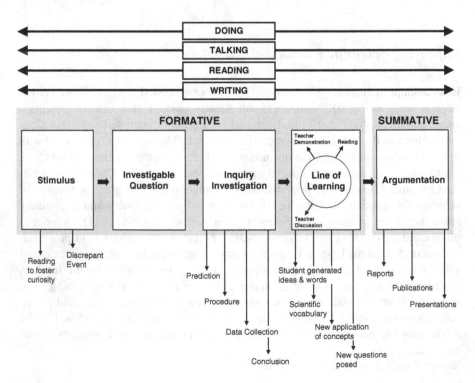

Fig. 5.1 Scientific literacy model (Adapted from Villanueva 2010)

resourced government schools whereby the members of the school community are isiXhosa first language speakers and English is the language of teaching and learning in the schools. The participating schools are representative of schools throughout the province where rote learning is prevalent, where learners very rarely engage in scientific investigations and higher-order writing activities, and science and English language proficiency are low. The schools were selected as a convenience sample in terms of an easily accessible cluster and principals' willingness to participate in the study. Schools within the cluster were then randomly allocated to either the experimental and control group. In the rural setting the sample included 168 grade 6 learners (*ns = 122 treatment; 46 control*) and seven teachers (*ns = 5 treatment, 2 control*) each from different schools. In the township setting, the sample comprised 675 grade 6 learners (*ns = 479 treatment; 196 control*) and eight teachers (*ns = 6 treatment, 2 control*) each from different schools.

Intervention, Data Collection and Analysis   At the onset of the study in each setting, the teachers from the treatment group participated in a two day professional development workshop on the theoretical and practical aspects of the scientific literacy model. They received the materials and resources used in the workshop, and the treatment teachers engaged in monthly onsite professional discussions and feedback received from the researchers. The control teachers continued with 'business as usual', e.g., their normal lessons throughout the period of the study.

In each setting, data were collected from both the teachers and learners. Teacher data was collected via four classroom observations and semi-structured interviews after each observation, and learner data via pre-post Raven's Standard Progressive Matrices (a nonverbal test used to measure reasoning ability), classroom observations, and a small sample of learners' notebooks.

The five-item *Science Notebook Checklist* assessed learners' writing in science and determined the degree to which their respective teachers guided and assisted learners to use inquiry skills and develop their procedural and conceptual knowledge in science. The checklist assessed the following five components: (1) constructing an investigable question, (2) designing an investigation, (3) collecting and recording data, (4) constructing scientific drawings, and (5) drawing conclusions. Each component was rated using levels 0–4. According to the instrument developers, the rating provides qualitative information to illustrate increased learner ownership of the respective component (Nesbit et al. 2004). More specifically, Level 0 indicated no evidence of the component present; Level 1 indicated that learners' writing was replicated directly from the teacher or text book; Level 2 suggests that the learner was able to generate his/her own information, but that some of the information written in the notebook contained misconceptions of the concepts; and Level 3 indicated that the learner generated his/her own ideas which were partially correct, yet omitted some important details. Finally, Level 4 suggests that the learner generated complete and accurate information. The data generated from the checklist were used to measure the level of learners' conceptual and procedural understanding when conducting scientific investigations. The researcher's observational notes were also used to supplement the findings from the checklist.

Summary of Science Notebook Findings

Constructing an Investigable Question The analyses of learners' science notebooks indicated that the majority of the learners from both groups copied their teacher's investigable questions throughout the term (level 1). The investigable questions that were analyzed in the learners' notebooks, particularly for the initial investigations, resulted from the activities modelled in the professional development workshops for teachers. While there was evidence that the teachers posed other questions, the questions were often closed-ended questions that did not provide opportunities for students to test their variables or develop a fair test. Questions commonly found in learners' notebooks included: "Which liquid is an acid or base?", "Which phase of water do you see?", or "Is [the animal] a vertebrate or an invertebrate?" Toward the middle of the term, more learners attempted to construct their own questions such as, "Will the object stick to the magnet?" In an investigation of electrical circuits, learners asked, "How can we make it bright?" as opposed to, "How can we make the bulb shine or glow brighter?" There was no evidence that learners constructed clear and/or accurate (level 4) investigable questions.

Designing an Investigation The learners from the urban and rural groups demonstrated evidence of an experimental procedure in their science notebooks. Most of the learners copied their teachers' sequential procedure (level 1), while nearly a quarter of the sample constructed and wrote plans for answering the question. Some learners' plans, however, omitted critical information of the design (level 2) and generally consisted of three or four incomplete sentences which were in some cases not written chronologically. Learners also displayed evidence of constructing an investigative plan, yet could have been enhanced by greater detail in their steps (level 3). Throughout the intervention, no learners showed evidence of being capable of writing a procedure which was complete and could be replicated (level 4; n=0).

Collecting and Recording Data Data from learners' science notebooks indicated that all learners collected and recorded data throughout the intervention. No learners scored in level 0 or level 1. The majority of learners were able to record data that was accurate, but could have been supplemented with additional information, such as consistently labelling units of measurement (level 3) and nearly one third of learners provided complete and accurate data in their science notebooks (level 4). Several learners produced level 2 entries, which contained inaccurate data. Examples of inaccurate data include utilizing the incorrect units of measurement, omitting key measurements or miscalculating averages or differences between several figures.

Scientific Drawings Results from the analysis of the scientific drawings illustrate a wide range of data in levels 1–4. This included: learners who copied teacher's drawings (level 1); and learners creating drawings which were labelled incorrectly or omitted relevant detail (level 2). Over half of the learners produced level 3 drawings, which, with more attention to labelling or the materials used, could advance to level

4. Several learners (n = 5) produced original drawings which were correctly labeled and provided details regarding their observations (level 4). Nearly all the learners dedicated half or three-quarters of the page to their scientific drawings and learners from the rural group demonstrated more drawings in levels 3 and 4.

Drawing Conclusions Low levels (levels 0–1) were prominent in both groups. Learners who scored at level 2 generally reiterated the results of the investigations with no connection to the target concepts of the lesson or unit. One quarter of the learners scored at levels 3 and 4.

Textual and Visual Representations

Overall, the experimental learners from the rural community progressed marginally in English writing, but improved considerably in isiXhosa. The improvements in isiXhosa writing were also evident in the urban setting, as was the learners' English writing ability. The development of learners' writing can be attributed to the embedded literacy and discursive practices that learners' were engaged with during their lessons. The learners had the opportunity to become immersed not only in English and isiXhosa, but also in the language of science.

The development of appropriate conclusions to the investigations appeared to be the most challenging step for learners to perform as a number of the student explanations were incorrect or excluded relevant details in the science notebook. The most common challenge was that some learners simply reiterated the steps of their procedure and repeated their data without critically analyzing the relationship between their predictions to the results of their investigation. The failure to evaluate empirical data against their initial assumptions was a missed opportunity for learners to reconstruct conceptual ideas and to understand the evaluative criteria used to establish scientific theories (Driver et al. 2000).

It is important to note that prior to the intervention the learners were rarely, if ever, encouraged to communicate their thoughts in writing. Consequently, the development of learners' writing skills in general appeared to be a difficult task for teachers when writing strategies were initiated. Before any writing could take place, teachers had to provide guidance and instruction on the literacy aspects of the science notebook approach, for example defining and clarifying scientific concepts such as 'procedure' and 'conclusion' in English and in isiXhosa. In addition, further instruction was required for understanding the scientific processes for each component. These observations of emerging writing and language skills are supported by the idea that oral language abilities and facilities with the mechanics of writing impact the quality and quantity of students' writing (Klein 2006). While some learners' writings reflected structural errors such as incomplete sentences, misspelled words or a lack of sequence of ideas, there was evidence of improvement in the way the students organized and recorded their findings. For the most part, the information in learners' notebooks displayed writings that were organized, could be used to make connections with prior experiences, and could be used as reference tool which learners could use to find evidence and support their thinking.

In addition to the inclusion of writing opportunities and practice in writing, another explanation for the learners' gradual improvements in writing could be credited to the method of writing that was promoted. With the introduction of science notebooks, learners were exposed to a structured form of writing and a systematic way of thinking. The science notebook framework assisted learners in developing a comprehensive understanding of process skills, as well as developing science concepts within the line of learning. As each entry began with an investigable question and ended with scientifically accepted ideas about the content, the process also allowed learners to engage in discursive practices and communicate scientific understandings based on their investigations.

With respect to the visual representations used in the science notebooks, learners' use of drawings was not heavily promoted in the classroom. For example, while instructing learners in data collection, one teacher from an urban school briefly stated, "You can also draw what you see" without further elaboration or follow-up. Yet, despite the lack of instruction to create visual representations nearly two-thirds of the notebook entries contained original drawings related to the investigation. Drawings which expressed the procedural aspects of the investigation were more prevalent with the rural participants. Learners from the rural school dedicated space in their notebooks for drawings, as opposed to merely drawing in the margins or within the text. Considering the general low use of English in the classroom, one may infer that the learners who were more inclined to communicate their ideas through pictures lacked the skill and relevant language necessary to communicate their experimental procedures. Research suggests that learners who have difficulty conveying their thoughts through text often exercise their artistic abilities as a means of communication (Dempster 2007). Using drawings or other forms of visual representations may be a vehicle for non-English speaking learners to scaffold their ideas, thus creating a connection between pictures, words, and concepts (Airey and Linder 2006; Lemke 1998). Although some of the drawings lacked pertinent details such as labelling, it seems that scientific drawings could be used as a springboard to develop language skills while simultaneously improving procedural understandings of investigations.

Alongside the visual representations, another important factor to consider when analyzing the learners' line of learning is learners' cognitive and linguistic abilities to communicate their ideas. The line of learning is an intensive language and cognitive process which requires that learners discuss and comprehend a range of science texts, contexts and multimodal representations such as the written word, symbols, formulae, diagrams and analogies. Based on the data generated from the analysis of learners' science notebooks, learners appeared to have struggled to process and develop their scientific and instructional language and to understand in English while reconciling these understandings with their home language of isiXhosa. While the learners' lack of English skills may have inhibited both their understanding and expression, what appears to be a problem with English-language development may, in fact, be a problem of concept development, or a combination of the two (Dempster 2007). Moodie (2009) suggests that where teachers fail to create experiences and activities for learners that illustrate or extend the concept, the learners' later lack of understanding can be misinterpreted as a language deficit. As such, the use of multiple modes of representation can be used to mediate these linguistic challenges.

In conclusion, the use of the science notebook improved over the course of the study, yet it appeared that learners' still struggled to engage in higher order writing tasks. It should be noted that the science notebook approach did not emphasize grammatical aspects, but was aimed at learners' formalizing their thinking in words. Learners often copied their teachers' writings from the blackboard or, alternatively, used bullet points or short incomplete sentences in their entries. Although writing-to-learn strategies in science are promoted to develop learners' scientific understandings, in the context of this study, learners' weak conceptual and procedural knowledge, as well as linguistic abilities, were contributing factors which influenced not only learners' writings (Gunel et al. 2007), but their ability to construct and communicate scientifically acceptable ideas.

Moving Forward in Multimodal Representation Research

As a science education research community, it is critical that we continue to question the theoretical and practical considerations of the classroom. While South African researchers and educators have cited practical barriers to achieving success in science classrooms, such as insufficient teaching and learning materials and shortages of competent science teachers, there appears to be other issues worthy of consideration. The review of the literature and the outcomes of the research presented in this chapter highlight the importance of engaging students in science discourse and practice, yet it also raises challenges regarding fulfilling the cognitive, epistemic and linguistic goals of school science and addressing the needs of the second-language learners. For instance, more empirical studies are required to understand how all learners engage in approaches to science learning that require the evaluation and communication of science in multiple modes. In other words, how are effective multimodal strategies conceptualized, used and aligned to support certain pedagogical approaches to learning science? And, possibly, whether or not the use of modes other than text can be utilized as a way to help students who are simultaneously dealing with the challenge of moving between languages and trying to understand science concepts and the processes of science?

This chapter highlighted research regarding ways in which learners' representations can be used as a starting point for learning. In addition, the ideas addressed can begin to support the idea that modes other than text can be used as bridges or anchors for instruction in science to help deal with some of the challenges of moving between different languages. The use of scaffolds and supports, opportunities to represent understandings in multimodal, modifications based on prior knowledge and experiences, systematic feedback, and non-threatening environments are aligned to theories of epistemological development which suggests that an individual's beliefs about knowledge and knowing may be derived and used according to epistemological resources that are environmentally contextual (Hammer and Elby 2002). As such, it may be beneficial for researchers to investigate what multimodal strategies or suite of strategies should be used or developed to assist teachers and students in science learning experiences. More examples which describe the successes and challenges of using multimodal representations are necessary to address these realities in South African science classrooms. As educators, it is critical that we: (1) understand students' mental and linguistic representations; (2) provide the space and time for students to solidify these representations with the language they are most proficient; (3) use multimodal representations to help bridge students' current representations and to develop their scientific ideas (which require claims, data & evidence), as well as their English language proficiency.

Implications for Teacher Professional Development

The studies presented in this chapter focused on the way in which second language English learners communicated their ideas via text. Although results from the studies cannot be generalised, the findings may provide some insight and constructive recommendations for science teacher development. It seems probable that pre-service and in-service training would benefit from integration of the intrinsic link between science and language and the ways in which language development is embedded in science instruction (Halliday and Martin 1993). In order to do this, teachers require a strong foundation in the cognitive academic language of science and the skills to help negotiate learners' everyday language and understandings to the language of science (Yore and Treagust 2006). Implicit in this idea, however, is the recognition that effective science teachers are required to have a good command of the discipline itself. Yore (2009) suggests that users of science discourse cannot fully comprehend the discourse without appropriate knowledge of the nature of science, scientific inquiry, and the content of science. As such science teacher training must emphasize both literacy aspects and a more comprehensive view of the scientific endeavor which includes an ability to communicate appropriately and accurately utilizing multiple modes of representation.

Moodie (2009), along with international researchers, contends that professional development programs should be ongoing and that "once-off" workshops so often offered to South African in-service teachers are probably grossly insufficient for improving their subject matter knowledge, pedagogical content knowledge, and issues of scientific literacy (Moodie 2009, p. 12). Learners' cognitive and language development is contingent on the opportunities teachers provide to express existing skills and to scaffold more complex ones via effective multimodal use. As such, professional development programmes and initiatives must acknowledge issues of the quality of prior training, as well as the amount of time and the attention required for teachers to acquire new skills and assimilate these approaches to their teaching environments (Hanley et al. 2007).

Being able to engage with science in a range of forms is an essential skill of a scientifically literate person (Crawford 2000; Fensham 2008; Norris and Phillips 2003; Yore and Treagust 2006). The ideas presented in this chapter beg the question of further investigation of the details of student-teacher interactions when facilitating dialogue, engaging in science and literacy activities, and constructing meanings with embedded representations.

References

Adelman, L. M., Falk, J. H., & James, S. (2000). Assessing the national aquarium in Baltimore's impact on visitor's conservation knowledge, attitudes and behaviors. *Curator, 43*(1), 33–62.

Airey, J., & Linder, C. (2006). *Languages, modality and disciplinary knowledge*. Paper presented at the ICLHE 2006: Integrating Content and Language in Higher Education, University of Maastricht. Maastricht, Netherlands.

Alant, B. (2004). Researching problem solving in introductory physics: Towards a new understanding of familiarity. *African Journal of Research in Mathematics, Science and Technology Education, 8*(1), 29–40. doi:10.1080/10288457.2004.10740558.

Albers, C., Rollnick, M., & Lubben, F. (2008). First year university students' understanding of validity in designing a physics experiment. *African Journal of Research in Mathematics, Science and Technology Education, 12*(1), 33–54. doi:10.1080/10288457.2008.10740627.

Braund, M., Ekron, C., & Moodley, T. (2013). Critical episodes in student teachers' science lessons using drama in grades 6 and 7. *African Journal of Research in Mathematics, Science and Technology Education, 17*(1–02), 4–13. doi:10.1080/10288457.2013.826966.

Christie, P., Butler, D., & Potterton, M. (2007). *Report of the Ministerial Committee Department of Education South Africa*. Retrieved March 5, 2008, from Independent Schools Association South Africa: http://www.isasa.org/component/option,com_docman/task,doc_download/gid,531

Crawford, B. (2000). Embracing the essence of inquiry: New roles for science teachers. *Journal of Research in Science Teaching, 37*(9), 916–937.

Cummings, W. K. (1984). Going overseas for higher education: the Asian experience. *Comparative Education Review, 28*, 241–257.

Dempster, E. R. (2007). Textual strategies for answering multiple choice questions among South African learners: What can we learn from TIMSS 2003? *African Journal of Research in Mathematics, Science and Technology Education, 11*(1), 47–60. doi:10.1080/10288457.2007.10740611.

Dempster, E. R., & Stears, M. (2013). Accessing students' knowledge in a context of linguistic and socioeconomic diversity: The case of internal human anatomy. *African Journal of Research in Mathematics, Science and Technology Education, 17*(3), 185–195. doi:10.1080/10288457.2013.839155.

Department of Arts, Culture, Science and Technology, South Africa. (1996). *White paper on science and technology*. Pretoria: Government Press.

Department of Science and Technology. (2007). *Ten year innovation plan*. Pretoria: Government Press.

Driver, R., Newton, P., & Osborne, J. (2000). Establishing the norms of scientific argumentation in classrooms. *Science Education, 84*, 287–312.

Dube, T., & Lubben, F. (2011). Swazi teachers' views on the use of cultural knowledge for integrating education for sustainable development into science teaching. *African Journal of*

Research in Mathematics, Science and Technology Education, 15(3), 68–83. doi:10.1080/1028 8457.2011.10740719.

England, P., Allison, P., Li, S., Mark, N., Thompson, J., Budig, M., & Sun, H. (2007). Why are some academic fields tipping toward female? The sex composition of U.S. fields of doctoral degree receipt, 1971–2002. *Sociology of Education, 80,* 23–42.

Falk, J. H. (2003). The museums and creativity pilot study. In G. Caban, C. Scott, J. Falk, & L. Dierking (Eds.), *Museums and creativity: A study into the role of museums in design education.* Sydney: Powerhouse Publishing.

Fensham, P. (2008). *Science education policy-making: Eleven emerging issues* (Section for science, technical and vocational education). Paris: UNESCO.

Fleisch, B. (2008). *Primary education in crisis: Why South African schoolchildren underachieve in reading and mathematics.* Cape Town: Juta & Co.

Gaigher, E., Rogan, J. M., & Braun, M. W. H. (2006). The effect of a structured problem solving strategy on performance in physics in disadvantaged South African schools. *African Journal of Research in Mathematics, Science and Technology Education, 10*(2), 15–26. doi:10.1080/1028 8457.2006.10740601.

Gee, J. P. (2002). Discourse and sociocultural studies in reading. In M. L. Kamil, P. B. Mosenthal, P. D. Pearson, & R. Barr (Eds.), *Methods of literacy research: The methodology chapters from the handbook of reading research* (Vol. III, pp. 119–131). Mahwah: Lawrence Erlbaum Associates.

Govender, N. (2007). Physics student teachers' mix of understandings of algebraic sign convention in vector-kinematics: A phenomenographic perspective. *African Journal of Research in Mathematics, Science and Technology Education, 11*(1), 61–73. doi:10.1080/10288457.2007. 10740612.

Govender, N. (2011). South African primary school teachers' scientific and indigenous conceptions of the earth-moon-sun system. *African Journal of Research in Mathematics, Science and Technology Education, 15*(2), 154–167. doi:10.1080/10288457.2011.10740709.

Green, W. J. (2007). Learner discourse and science learning in the context of microcomputer-based laboratory (MBL) collaborative learning activities. *African Journal of Research in Mathematics, Science and Technology Education, 11*(1), 1–16. doi:10.1080/10288457.2007.10740607.

Gunel, M., Hand, B., & Prain, V. (2007). Writing for learning in science: A secondary analysis of six studies. *International Journal for Science and Mathematics Education, 5*(4), 615–637.

Halliday, M., & Martin, J. (1993). *Writing science: Literacy and discursive power.* Pittsburgh: University of Pittsburgh Press.

Hammer, D., & Elby, A. (2002). On the form of a personal epistemology. In B. K. Hofer, & P. R. Pintrich (Eds.), *Personal epistemolgy: The psychology of beliefs about knowledge and knowing* (pp. 169–190). Mahwah: Lawrence Erlbaum.

Hanley, N., Wright, E. R., & Alvarez-Farizo, B. (2007). Estimating the economic value of improvements in river ecology using choice experiments: An application to the water framework directive. In S. Navrud & R. Ready (Eds.), *Environmental value transfer: Issues and methods* (pp. 111–130). Amsterdam: Springer.

Heugh, K. (2000). *The case against bilingual and multilingual education in South Africa* (PRAESA Occasional Paper No. 6). Cape Town: PRAESA.

Heugh, K. (2013). Multilingual education policy in South Africa constrained by theoretical and historical disconnections. *Annual Review of Applied Linguistics, 33,* 215–237.

Holmarscottir, H. (2003, December). *Literacy in two languages? Implementation of language policy in a post-colonial context.* Paper presented at the Neetreed conference, Gausdal Hoyfjellshotell, Norway.

Howie, S., & Plomp, T. (2005). International comparative studies of education and large scale change. In N. Bascia, A. Cummings, A. Datnow, K. Leithwood, & D. Livingstone (Eds.), *International handbook of educational policy* (pp. 75–100). Dordrecht: Kluwer Press.

Kallaway, P. (2007). The profound crisis of teaching. Retrieved January 19, 2008, from Mail and Guardian Online: http://www.mg.co.za/article/2007-08-21-the-profound-crisis-of-teaching

Klein, P. D. (2006). The challenges of scientific literacy: From the viewpoint of second-generation cognitive science. *International Journal of Science Education, 28*(2–3), 143–178.

Klein, P. D., & Kirkpatrick, L. C. (2010). Multimodal literacies in science: currency, coherence and focus. *Research in Science Education, 40*(1), 87–92.

Lelliott, A. (2010). The concept of spatial scale in astronomy addressed by an informal learning environment. *African Journal of Research in Mathematics, Science and Technology Education, 14*(3), 20–33. doi:10.1080/10288457.2010.10740689.

Lemke, J. L. (1998). Resources for attitudinal meaning: Evaluative orientations in text semantics. *Functions of Language, 5*(1), 33–56.

Lemke, J. L. (2000). Across the scales of time: Artifacts, activities, and meanings in ecosocial systems. *Mind, Culture, and Activity, 7*(4), 273–290.

Lemmer, M. (2011). Analysis of South African grade 10 learners' conceptual resources regarding the concept of energy in physics. *African Journal of Research in Mathematics, Science and Technology Education, 15*(1), 4–17. doi:10.1080/10288457.2011.10740698.

Mammino, L. (2001). Physical quantities and their changes: Difficulties and perceptions by chemistry students. *African Journal of Research in Mathematics, Science and Technology Education, 5*(1), 29–40. doi:10.1080/10288457.2001.10756127.

Mayaba, N., Otterup, T., & Webb, P. (2013). Writing in science classrooms: Some case studies in South African and Swedish second-language classrooms. *African Journal of Research in Mathematics, Science and Technology Education, 17*(1–02), 74–82. doi:10.1080/10288457.2013.826972.

Mbano, N. (2004). Pupils' thinking whilst designing an investigation. *African Journal of Research in Mathematics, Science and Technology Education, 8*(2), 105–115. doi:10.1080/10288457.2004.10740565.

Meiring, L., & Webb, P. (2012). The effect of toys as educative material on pre-service education students' understanding of energy. *African Journal of Research in Mathematics, Science and Technology Education, 16*(1), 100–111. doi:10.1080/10288457.2012.10740732.

Moodie, P. (2009). *A response to the intermediate phase recommendations of the implementation review panel, from a natural sciences and technology perspective.* Johannesburg: Setlhare Science Curriculum Trust.

Nesbit, C., Hargrove, T., Harrelson, L., & Maxey, B. (2004). Implementing science notebooks in primary grades. *Science Activities Journal, 40*(4), 21–29.

Norris, S., & Phillips, L. (2003). How literacy in its fundamental sense is central to scientific literacy. *Science Education, 87*(2), 224–240.

Probyn, M. J. (2001). Teachers' voices: Teachers' reflections on learning and teaching through the medium of English as a second language. *International Journal of Bilingual Education and Bilingualism, 4*(4), 249–266.

Probyn, E. (2004). *Blush: Faces of shame.* Minneapolis: University of Minnesota Press.

Qhobela, M., & Moru, E. K. (2011). Learning physics through argumentation at secondary school level in Lesotho: A feasible teaching strategy? *African Journal of Research in Mathematics, Science and Technology Education, 15*(2), 205–220. doi:10.1080/10288457.2011.10740713.

Ramorogo, G., & Ogunniyi, M. (2010). Exploring teachers' conceptions of the rainbow using an argumentation-based intervention. *African Journal of Research in Mathematics, Science and Technology Education, 14*(1), 24–35. doi:10.1080/10288457.2010.10740670.

Reddy, V. (2006). *Mathematics and science achievement at South African schools in TIMSS 2003.* Cape Town: HSRC Press.

Reddy, V., Juan, A., Gastrow, M., & Bantwini, B. (2009). *Science and the publics: A review of public understandings of science studies.* Cape Town: HSRC Press.

Setati, M., Adler, J., Reed, Y., & Bapoo, A. (2002). Incomplete journeys: Code-switching and other language practices in mathematics, science and English language classrooms in South Africa. *Language and Education, 16*(2), 128–149.

Statistics South Africa. (2011). *Consumer price index: October 2011, statistical release P0141.* Pretoria: Statistics South Africa.

Stears, M., Malcolm, C., & Kowlas, L. (2003). Making use of everyday knowledge in the science classroom. *African Journal of Research in Mathematics, Science and Technology Education,* 7(1), 109–118. doi:10.1080/10288457.2003.10740553.

Tang, K. S., & Moje, E. B. (2010). Relating multimodal representations to the literacies of science. *Research in Science Education, 40*(1), 81–85.

Tang, K. S., Delgado, C., & Moje, E. B. (2014). An integrative framework for the analysis of multiple and multimodal representations for meaning-making in science education. *Science Education, 98,* 305–326.

Taylor, N., & Vinjevold, P. (1999). Getting learning right (Report of the Presidents' Education Initiative Research Project). Johannesburg: Joint Educational Trust.

Villanueva, M. G. (2010). *Integrated teaching strategies model for improved scientific literacy in second-language learners.* Unpublished Ph.D. Thesis. Nelson Mandela Metropolitan University. South Africa.

Villanueva, M. G., & Webb, P. (2008). Scientific investigations: The effect of the 'science notebooks' approach in grade 6 classrooms in Port Elizabeth, South Africa. *African Journal of Research in Mathematics, Science and Technology Education, 12*(2), 3–16. doi:10.1080/10288 457.2008.10740631.

Webb, P., & Mayaba, N. (2010). The effect of an integrated strategies approach to promoting scientific literacy on grade 6 and 7 learners' general literacy skills. *African Journal of Research in Mathematics, Science and Technology Education, 14*(3), 34–49. doi:10.1080/10288457.2010. 10740690.

Webb, L., & Webb, P. (2008). Introducing discussion into multilingual mathematics classrooms: An issue of code switching? *Pythagoras: Teaching and learning mathematics in multilingual classrooms: Special Issue, 67,* 26–32.

Webb, P., Williams, Y., & Meiring, L. (2008). Concept cartoons and writing frames: Developing argumentation in South African science classrooms? *African Journal of Research in Mathematics, Science and Technology Education, 12*(1), 5–17. doi:10.1080/10288457.2008.1 0740625.

Wessels, H., Wessels, D., & Nieuwoudt, H. (2006). Arrangement strategies in data representation tasks of grade 4–7 learners. *African Journal of Research in Mathematics, Science and Technology Education, 10*(2), 39–48. doi:10.1080/10288457.2006.10740603.

Yore, L. (2008). Scientific literacy for all students: Language, culture, and knowledge about nature and naturally occurring events. *L1 – Educational Studies in Language and Literature, 8*(1), 5–21.

Yore, L., & Hand, B. (2010). Epilogue: Plotting a research agenda for multiple representations, multiple modality, and multimodal representational competency. *Research in Science Education, 40*(1), 93–101. doi:10.1007/s11165-009-9160-y.

Yore, L., & Treagust, D. (2006). Current realities and future possibilities: Language and science literacy- empowering research and informing instruction. *International Journal of Science Education, 28*(2–3), 291–314.

Zimmerman, L., Howie, S., & du Toit, C. (2009). *Comprehending the macro through the lens of the micro: Investigating South African Grade 4 teachers' reading literacy instruction practices.* Paper presented at the Teacher Education Research and Development Programme (TEP) Conference 2008 (pp. 31–44). Braamfontein: Centre for Education Policy Development (CEPD).

Chapter 6
Scientific Concepts, Multiple Modalities, and Young Children

Deborah L. Linebarger and Lori Norton-Meier

Learning represents an active and constructive process whereby children engage with, make sense of, integrate, and encode new knowledge with existing knowledge (Kozma 1994). During this process, there is an interaction among a learner's individual differences, goals or motivations, abilities, knowledge, and developmental capacities and constraints. To develop competence in a particular conceptual domain, children must possess a deep foundation of factual knowledge, develop a framework that represents this knowledge base appropriately, and organize or represent this knowledge within the framework in ways that maximize its later retrieval and application (Bransford and Schwartz 1999). Experts in particular areas not only have extensive knowledge of that area but also notice, organize, represent, and interpret information associated with this area in ways that are fundamentally different when compared with novice learners (Bransford and Schwartz 1999). While children differ from adult learners in many ways, there are a number of commonalities that, when supported within a learning environment, can facilitate their competence within and across various domains. Research and recent syntheses of learning during early childhood indicate that very young children are capable of engaging in sophisticated cognitive activities well before they enter formal schooling (Gopnik 2012). For instance, infants attend to regularities or statistical patterns in a variety of sensory stimuli including speech patterns as they begin to make sense of everyday experience (Lany and Saffran 2013). During early childhood, young children spontaneously invent experiments in their play (Cook et al. 2011) as they explore why an event happens, whether it always happens, and under what conditions it

D.L. Linebarger (✉)
University of Iowa, Iowa City, IA, USA
e-mail: deborah-linebarger@uiowa.edu

L. Norton-Meier
University of Louisville, Louisville, KY, USA
e-mail: lori.nortonmeier@louisville.edu

© Springer International Publishing Switzerland 2016 97
B. Hand et al. (eds.), *Using Multimodal Representations to Support Learning in the Science Classroom*, DOI 10.1007/978-3-319-16450-2_6

might vary. Imagine an infant in a high chair who drops her spoon to the floor. Her parent will pick it up and place it back on the tray. The infant will likely drop it again to see if the spoon falls like before. Once assured that it does, she will start dropping everything off her tray, varying the conditions as she comes to understand gravity.

Both infant pattern detection and experimentation are inherently social (Gopnik 2012). Young children learn by observing others' experiences (Bandura 2001) both directly and indirectly (e.g., via screen media) as well as by interacting with others. Research confirms that more positive peer play in preschool is linked to better emotion regulation, higher levels of autonomy, and larger receptive vocabularies as well as lower rates of aggression, shyness, and adjustment difficulties (Fantuzzo et al. 2004). Engaging in dialogue around learning provides opportunities for young children to use rich and sophisticated vocabulary with peers while they negotiate their learning and understanding (Teale and Sulzby 1986). Conversations require that children receive linguistic input from peers, translate that into a mental representation, and then manipulate and reproduce that representation as they continue the peer conversation (Conezio and French 2002).

Mental Representation

These early cognitive and social activities (i.e., pattern detection, experimentation, and social interactions) are essential for helping very young children accurately represent their everyday experiences as well as acquire new and varied processes and strategies to manipulate and later reproduce these representations. Mental representation refers to the cognitive process through which knowledge is acquired, organized, and subsequently made available for use (Sigel 1999). Representational processing requires young children to transform incoming perceptual and sensory experiences into a mental representation of these experiences that can be understood, manipulated, and reproduced (Damon et al. 1998; Sigel 1999). Children construct multiple representations of various concepts by repeatedly engaging with the same referent across multiple forms of expression. For instance, children's understanding of a cup could include their direct experience with a cup in their home, a cup they saw in a photograph in a book, and a cup they saw a favorite TV character using onscreen. In all three situations, the cup is equivalent in meaning; however, it differs in form of expression. Our minds are designed to work in a cross-modal way; that is, as we gain experience and construct knowledge about our concept of cup, we shuttle information about cup back and forth from one mode of representation to another. When each shuttle occurs, it is transformational in the sense that shifting among modes creates new meanings and understandings about our general concept of cup. Ultimately, no single mode for representing knowledge is powerful enough to explain or capture the perceptual and sensory information associated with cup. Consequently, children must develop and exercise considerable representational flexibility as they learn.

Representational Flexibility

Representational flexibility refers to the ability to easily shift between different modes of expression in order to perceive, process, manipulate, act on, and repro-duce knowledge (Hayne 2006). In early childhood, representational flexibility is supported by multiple opportunities to encode information across a variety of con-texts ultimately contributing to the formation of hierarchical and relational repre-sentations of events and experiences compared with encoding only specific attributes of an event or experience. The consequence of encoding specific attributes is that later retrieval and subsequent transfer of these attributes is limited to situations where there is a direct match between the original learning context and the current situation (see Barr and Brito 2013; Richmond and Nelson 2007 for reviews). As representational flexibility develops during this developmental period, young chil-dren will become better able to comprehend and learn from novel situations that, although not identical, do possess similarities with the original stored representation and the current situation (Barr and Brito 2013).

Modes of Expression

There are multiple forms of expression or modes through which information is con-veyed. These modes can be used as generative tools when young children are con-structing and reproducing their knowledge about particular concepts (Yelland et al. 2008). Prior to formal schooling, very young children are immersed with and inter-acting with a variety of modes or symbol systems including aural (e.g., music), visual (e.g., pictures, art), aural/visual (e.g., television, video games), linguistic (e.g., storytelling), gestural (e.g., body control, emotion), and spatial (e.g., geo-graphical, architectural) modes. They become quite proficient at interpreting and subsequently using these modes to make sense of their worlds. For instance, very young children will frequently draw pictures that reflect their mental representa-tions about various objects, events, or persons. Figure 6.1 depicts a 2-year-old child's mental representation of a car. As adults, we examine the picture and see a mass of scribbles. The little boy who drew the picture also narrated while drawing stating that his car went "vroom, vroom" as he scribbled furiously in the highlighted area. All of the scribbles represented his current understanding of a car including how it moved through space, sped up, or slowed down. In this example, the boy was using a variety of different modes to represent his knowledge of a car including linguistic (as he narrated), visual (the picture itself), and gestural (the varying move-ments of his hand/arm as he sped up and slowed down) modes.

Fig. 6.1 A 2-year-old
child's mental
representation of a car

Implications of Early Childhood Cognition for Science Instruction

Once in school, there is a decided shift away from the use of multiple modes as forms of expression and knowledge construction to representation that unilaterally privileges print-based text (Kress 1997). By emphasizing a unimodal world, educators are losing opportunities to maintain the rich, multimodal worlds experienced in early childhood. Instead, mismatches between the multiple ways that young children take in information from the environment, manipulate that information, play with and make sense of it, and subsequently develop a sense of self including how to communicate with the broader world about this information (Pahl and Rowsell 2006; Yelland et al. 2008) and the ways in which they are now asked to understand and convey information occur narrowly shape how information is perceived and stored and likely limit how it can be used to further understanding. As a consequence, early elementary school instruction in science leads to less risk-tasking, fewer innovations, and more emphasis on creating a unimodal and relatively fixed knowledge base about the world (Kress 1997).

In addition, young children often have few opportunities to engage in the practices of acquiring, evaluating, and integrating information from multiple sources and multiple modes of presentation. These practices exert considerable demand on young children's cognitive processes; however, such demand is necessary for success in science. Retrospective interviews with adult scientists indicated that their interest in science began early and was often linked to experiences that inspired them or that they could identify with (e.g., how scientists were portrayed in science fiction texts; O'Keeffe 2010). These scientists were more likely to identify informal science learning experiences as instrumental in their desire to pursue science careers when compared to formal learning experiences. As such, creating early childhood environments that match how young children prefer to and naturally do learn could play a large role in not only scientific achievement but also in keeping young children engaged in scientific pursuits as they proceed through school and into their careers.

The Science Writing Heuristic (SWH) Approach

One intervention that holds promise in both supporting and more importantly sustaining young children's early representational flexibility and their acquisition and use of science knowledge is the Science Writing Heuristic (SWH) approach. The SWH approach was developed by Hand and Keys (1999) as a language based approach to science inquiry. Importantly, the approach recognizes that science cannot be done without language (Norris and Phillips 2003), and that the learner must be actively involved in the inquiry process. Students pose questions, design activities, gather data, generate claims, and derive evidence from the data. This process requires students to publicly and privately negotiate the science ideas that are being explored through the inquiry activities. Importantly, the science concepts are not taught separately from the inquiry activities. Students learn the science concepts and then do activities to confirm their learning and understanding. In this approach the students are learning both the science concepts alongside the practices of science (Ford and Forman 2006).

Much of the research investigating the efficacy of the SWH approach has been conducted with upper elementary, middle and high school children and teachers (e.g., Akkus et al. 2007) and at the tertiary level (e.g., Burke et al. 2006). These studies document significant science learning gains. Importantly, results also indicate significant critical thinking gains that are beyond specific science learning gains for students who participate in this approach.

Initial research with lower elementary school students (e.g., K-3rd grade) details that the SWH approach can be successfully implemented in these grades and that, with instruction, young students are able to engage with the basic concepts of inquiry. Sanders (2011), in her work at the K-1 level, writes that "the SWH approach encourages students to use a variety of resources including: contacting professionals in the specific area of study, using various media materials, and partaking in field trips, generating assorted experiments, and using inventive methods of presenting learned information such as poster presentations, reader's theatre scripts, poetry, dioramas" (p. 73). Another study found that kindergarten students involved in the SWH environment generated, with help from a musician, a song about trees that reflected the big idea they were studying (Nelson 2011). A further study by Malin (2011) showed how students in kindergarten and grade 1 were able to construct class concept maps using large sheets of paper and post-it notes that were then successfully transferred to more individually focused concept maps. These forms of representation offer evidence that these young students are able to move beyond normal and individual modes of presentation provided by teachers and to focus successfully on creating combinations of visual and text representations.

The previous examples of work with students in K – 3rd grade indicate what is possible, Johnson (2011) examined the organizational structures that are required to achieve success within SWH classrooms. She suggested five classroom environmental conditions were required for success: student-centered learning, relationships and culture for learning, organization and management, physical arrangement, and interactions. While each are these are critical factors in the structure of the environment(s) needed for success in using the SWH approach, they do not highlight the type of process factors in the environment that are essential to the success of the SWH approach. These process factors result from the language environment created within an SWH classroom that ultimately enables students to learn about language by using language as they live language during scientific inquiry. Understanding the interplay between language and science learning is critical to the foundation of the SWH approach.

Language and Science Learning

Recent pressures to make younger grades and preschool more academic and structured are contrary to how young children learn (Gopnik 2012; Weisberg et al. 2013). For example, in science classes, explicit teaching can narrow the range of hypotheses children will consider because teachers are considered powerful sources of information. Consequently, young children will quickly default to any information that teachers offer and limit their focus on the range of hypotheses or possibilities presented to them. The SWH, on the other hand, is based on children generating the questions, making claims and providing evidence for those claims and then negotiating their evidence with peers without the explicit instruction of their teachers. This process is quite similar to the patterns of learning and play young children engaged in prior to formal schooling although formal schooling plays a critical role in helping children structure the process as they continue to acquire the skills and practices of science.

Questions regarding the characteristics of science learning environments that should be provided for young children are demanding greater attention recently. Much of the early research was framed around a deficit model of education where findings were collected about what young learners *lack* and how teaching should be conducted to overcome these deficits. More recently, researchers are arguing for an additive model. Young learners are capable of sophisticated cognitive work that can be supported, sustained, and further developed when the classroom environment and the language used within that environment can leverage these young children's developmental and cognitive predispositions and strengths.

One theoretical framework about language use conceptualized broadly across multiple modes of expression including text, picture, diagrams, audio-visual content, etc. suggests that students must learn *about* language prior to them effectively and efficiently using language in the context of learning (Halliday and Martin 1993). This view suggests that language is initially decontextualized and, consequently, learning how to use the language occurs first in order to use the language for learning. An alternative framework is that language as a tool must be learned but also that, through the process of using language as a learning tool, children will come to know more about the topic than was known prior to its use. In this view, language becomes an epistemological tool that not only results in the generation of some product (i.e., knowledge) but, as a consequence of being engaged in the process of production, also results in a richer understanding of the content. Underpinning this framework is the early work of Halliday (1975) who proposed that students learn *about* language, *through* using the language as they *live* the language. Particular emphasis is placed on the process of using language in the context of learning about science. Both the process and the product (i.e., language and science knowledge, respectively) are inseparable.

Norton-Meier's (2008) model of embedded language practices highlights the interaction between and among each of these elements, illustrating how argument-based inquiry can be developed and implemented through a language lens (Fig. 6.2).

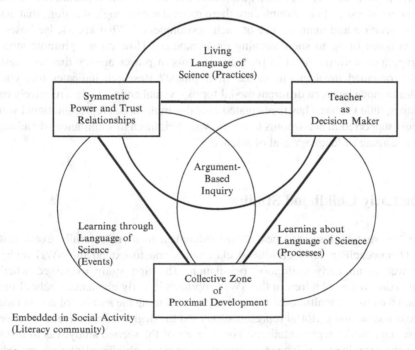

Fig. 6.2 A model to examine embedded language practices within argument-based classrooms

Three features of SWH classrooms influence the environment in which inquiry occurs and the impact such an environment has on both teacher practices and child outcomes. These three features are: (1) teacher as decision maker, (2) symmetric power and trust relationships, and (3) the collective zone of proximal development. Each of these features is fundamentally linked to the construction and maintenance of language-based interactions within classrooms that adopt this language-as-learning-tool approach. The model highlights the intersecting nature of these particular language opportunities and how these elements can frame the learning environments that are critical for student success. The three critical elements are defined below (based on Norton-Meier 2008)

> Teacher as Decision-Maker: If the teacher is viewed as a trusted curricular decision maker who is competent to decide—based on the students' learning needs as well as their questions—what to do next, the teacher has greater success at implementing the SWH in the classroom.
>
> Symmetric Power and Trust Relationships: The teachers recognize that their students control learning and with that in mind, what then, is the teacher's role in the classroom? Teachers work to find ways to validate student questions and support individual students in their own science inquiry.
>
> Collective Zone of Proximal Development: The collective zone of proximal development in SWH classrooms is that space between where a learner can go alone and where one can go with others learning as a group. The teacher creates a space or a zone of functioning for a group in which learning is supported through guidance and participation in the classroom community.

There are a number of critical questions that remain unanswered about how these elements intersect. For example, are there particular pedagogical actions that assist the generation and maintenance of such environments? What knowledge bases do the children bring to such learning environments? How do we promote student engagement with the need to provide elements of power agency that will ensure their continued involvement in the environment? Research indicates that young children move between different modal forms – visual and text – in a relatively easy manner, little research has investigated how this multimodal representational world is best supported in classrooms to enable young children to simultaneously advance their learning of language and of science.

The Early Childhood Studies

The two studies described next were undertaken to more carefully examine the SWH approach in early childhood classrooms and to extend the SWH body of research to an early childhood population. The first study examined whether immersing young children in the SWH approach in early elementary school influenced their use of multimodal representations including the number of content ideas across textual and pictorial representations and the degree of cohesiveness between these multimodal representations. The purpose of the second study was to investigate the complexity of language that arises in early childhood classrooms while young children engage in science inquiry with embedded language practices (e.g., providing opportunities for children to engage in reading, writing, speaking, listening, visualizing, and viewing within the context of science learning). Considering the two studies together begins to explore the relationship between an argument-based, language embedded science learning environment and student multimodal communication. If, as the research discussed previously suggests, early childhood learning environments would be more effective if they included multimodal communication opportunities and if the SWH provides an instructional avenue for creating effective multimodal learning environments, then work clarifying the characteristics of effective SWH learning environments is needed. The studies presented here begin to explore the relationship between the characteristics of SWH learning environments and student multimodal communication.

Study 1: Early Childhood, Mental Representation, and Cognitive Flexibility

Many students, especially younger learners, have few opportunities to engage in the practices of acquiring, evaluating, and integrating information from multiple sources and multiple forms of presentation. Lemke (1994) states, "what it means to be able to use a scientific concept, and therefore to understand it in the way that a scientist does, is to be able to fluently juggle with its verbal, mathematical, and visual/graphical aspects, applying whichever is most appropriate in the moment and freely translating back and forth among them." For students, this ability to translate these multiple modes of representation places a high demand on their cognitive processes. Students not only need to understand and interpret content within a modal representation but also must be able to translate between modes when synthesizing a written composition (Hand et al. 2014). Helping students learn how to create effective representations is not a straightforward exercise (Danish and Phelps 2011). Yoon (2012) argued that there is a critical need to create learning environments where students can engage in diverse opportunities to use *representational resources* with *appropriate demands* that encourage them to represent and constitute their understandings in multiple ways beyond only written texts. With this in mind, writing samples were collected from two groups of children in grades 1 through 3: (a) children in SWH classrooms where they were immersed in language-rich learning environments in science classrooms and (b) children in non-SWH classrooms.

Participants In one school district in a small mid-western community, 19 teachers of 369 students in 1st, 2nd, and 3rd grade classrooms either delivered science instruction through a traditional science curriculum approach (n = 8 teachers, 160 students) or by using the SWH approach (n = 11 teachers, 209 students).

Measure Near the end of the school year, students were asked by their teachers to generate a writing sample that described their conceptual understanding of a specific topic (e.g., plant growth, the solar system) by incorporating different modes of expression about the topic including printed text and pictures. A coding scheme was developed to analyze multimodal representations and representational flexibility.

Coding Scheme SWH is fundamentally linked to representational flexibility. Children must be able to flexibly shift among varied representational modes that appropriately and accurately convey their understanding of particular concepts. A first step in examining representational flexibility involved coding the number of content ideas conveyed via printed text and via pictures to determine whether and how SWH children differed in their production of textual and pictorial representations (i.e., modes of expression). Within each representational type, we examined whether the representation was linked to the key scientific concept being described (i.e., representational flexibility).

Second, we created codes that examined the degree of cohesiveness between children's text and pictorial representations. Cohesiveness was defined as the level of integration between text and pictorial content and was categorized into either low cohesiveness (i.e., used either text or pictures or, if 2 modes were used, made no connections between the text and the picture) or high cohesiveness (i.e., 2 or more modes were used, the content depicted via each mode was connected and well-integrated).

Results

Modes of Expression For text content, a slightly larger (although significant) percentage of SWH children (99.5 %) linked their text representations to the key scientific concept when compared with non-SWH children (95.6 %). The difference was substantially larger for picture content. Specifically, 66.4 % of SWH children linked their picture representations to the key scientific concept compared with just 33.6 % of non-SWH children. SWH children's picture representations were also better differentiated (64.7 %) when compared to their non-SWH peers (35.3 %). Differentiation refers to their ability to include specific concepts in their drawings. For instance, SWH children's drawings illustrating their knowledge of plants included specific elements related to plant growth (e.g., roots, leaves, soil, sun, water).

Representational Flexibility Representational flexibility and multimodal use between children in SWH classrooms and those in non-SWH classrooms was examined by contrasting the percentage of children in each group who included both text and picture representations that were directly linked to the main scientific concepts they were presenting. Only 65.9 % of non-SWH children were able to link both their text and their picture representations to the main scientific concept whereas 92.3 % of SWH children linked both.

Cohesiveness SWH children's writing samples were significantly more likely to be categorized as highly cohesive when compared with non-SWH children's writing samples: 62.3 % in SWH classrooms versus 37.7 % in non-SWH classrooms.

Study 1: Conclusion and Implications

Immersion in the language-rich environments created by teachers using the SWH approach likely promoted the development of coherent conceptual understandings. Further, SWH holds promise as an intervention that supports and maintains young children's previous facility with multiple modes of expression. Future research includes an expansion of the coding system to further examine cohesiveness and modes of expression as well as an examination longitudinally of how cohesiveness, multimodality, and scientific understanding evolve during early childhood. These results underscore, however, that very young children are capable of engaging in profound cognitive work when immersed in a world of rich and varied modes of expression (Gopnik 2012; Kress 1997). As young children transition into formal schooling, their exposure to an environment that continues to foster scientific induction often disappears in favor of a more didactic approach centered on traditional curricular methods. The SWH approach offers an alternative to these traditional curricular approaches that is consistent with and, therefore, supports and maintains a child's natural cognitive capabilities and tendencies including the use of mental representations and the development of representational flexibility as well as the focus on the use of multiple modes of expression for both processing and producing scientific knowledge. Our research provides initial evidence that SWH is an effective approach in fostering representational flexibility and cohesiveness across multiple modes of expression. Our future research will continue to explore these concepts as they relate to early childhood cognitive development and as they predict later academic success.

Study 2: Teachers and Language in Early Childhood Classroom Environments

In order to study how teachers engage with a language environment as part of implementing the SWH approach, teacher implementation of the approach over an extended period of time was needed. The process of change in instructional style is a difficult one for teachers. In Study 1 described above, children's ability to cohesively integrate text and picture was directly and powerfully tied to their teacher's years of experience implementing SWH (Aguirre-Mendez et al. 2013). In fact, students whose teachers had less than one year of experience implementing SWH were 1.28 times *less* likely to demonstrate high cohesiveness. In contrast, students whose teachers had at least a year of experience but fewer than 2 years were 2.08 times *more* likely to demonstrate high cohesiveness. The effect was particularly large for students whose teachers had 2 or more years of experience; they were 6.27 times *more* likely to demonstrate high cohesiveness between text and picture representations. Understanding exactly what kinds of environments highly experienced teachers create and how these environments interact with young children's representational capacities is the focus of Study 2: The Teachers' Study.

The purpose of Study 2 was to investigate the complexity of language that arises in early childhood classrooms while young children engage in science inquiry with embedded language practices. This type of thematic-based science literacy curriculum unit is an innovative strategy that successfully embeds language practices into the classroom and contributes to the young child's developing literacy capabilities using a focus on science inquiry.

Participants Early childhood teachers (K-3) were involved in a summer workshop, implementation and analysis cycle for 3 years from 3 school districts. The distribution of school districts allowed for a diverse range of contexts and populations including a small population of students who were English language learners (n = 14). It was originally proposed that 20 teachers would participate in the project. After Year 1, the interest by other teachers and administrators in the same school increased. Twenty-nine teachers participated in the project for 2 years or 3 years (28 female, 1 male) representing 2–25 years of teaching experience. The numbers shifted each year based on natural attrition factors found in the realities of school environments (e.g., change in grade level assignment, family reasons, or the teacher moving out of the district or participating school). For the purposes of this chapter we chose to only analyze the data for teachers who participated in the project 2 or 3 years.

Procedure The teachers were invited to attend a summer institute at Iowa State University for 5 days to participate in a science content update, critical reading experiences, and science inquiry pedagogical strategies with embedded language practices. Teachers also participated in three follow-up days in their own school districts where, with the assistance of project staff, they worked together during the school year planning SWH units, implementing the curriculum in the classroom, and contributing to ongoing data collection and analysis. This pattern of five summer in-service days and 3 school year follow-up days continued each year of the 3-year study. Teachers were asked to implement 2 SWH units in Year 1 (one in the fall, one in the spring), 3 units in Year 2, and 4 units (in most cases, the entire science curriculum) in Year 3.

Data Sources There were five sources of data collected: Observations – these were scored using a modified Reform Teacher Observation Protocol (Sawada et al. 2002); Teacher Interviews; Pre-Post test/interviews with children; Writing samples from each SWH unit; and Collection of other artifacts. A team of 3 raters was employed to score the teacher observations as per the modified RTOP, code the lessons for language events based on the model discussed above and to code the writing samples.

From the coding of all the language events that occurred in the teachers' classrooms from those that participated for the 3 years of the project, a series of codes were generated. These codes were then matched against the theoretical model described above to determine an overall perspective of how language was being engaged with in these early elementary classrooms. This lead to the following codes displayed in Fig. 6.3.

Codes	Teacher as Decision-Maker (big ideas-unit plan)	Symmetric Power & Trust Relationships (control of learning)	Collective ZPDs Mediation - - Argument (dialogic interactions)
About Language	Just in Time-about (JiT-a) Writing – technical, form – (Wi-a)	Dialogue (Di)	Science Argumentation/ language (SAL) Question, Claim, Evidence
Through Language	Just in Time-through (JiT-t) Writing to learn (Wi-t)	Change in Knowledge (CiK) Way of Knowing (WoK)	Border X (BX)
Live the Language	Just in Time-live (JiT-a) Writing as a scientist (Wi-L)	Access (Ac) Funds of Knowledge (FoK)	Reading as act of inquiry (RasI)

Code descriptions:

Just in Time – About: Teacher provides just in time instruction about the technical aspects about language (structure, grammar, spelling, conventions, and the like)

Writing – About: Writing shared or discussed about how writing works particularly focused on technical aspects of writing such as form or structure.

Dialogue: Students engage in dialogue that is among students and not teacher-controlled discourse.

Science Argumentation Language: The language of argument such as questions, claims, evidence.

Just in Time – Through: Teacher provides just in time instruction about learning through language, specifically learning through reading and writing to understand science concepts.

Writing to Learn – Through: Writing experiences that allow students to write to learn or learn through writing particularly about their inquiries in multimodal means (for example in a science notebook).

Change in Knowledge/Ways of Knowing: Teachers and students recognizing when they have a change in understanding or discussing their ways of understanding new content (for example, students discussing that they understand the weather through rural farming practices).

Border X: Teacher helps students to navigate the borders between disciplines (literacy and science), academic language vs. every day language, text to text, unit to unit, home to school.

Just in Time – Live: Teacher creates just in time instruction that helps students to experience living the language of science and the way scientists use language.

Writing as a Scientist – Live: Opportunities for the students to write like scientists and focus on communicating with different audiences about their findings and developing understanding.

Access/Funds of Knowledge: Teachers help students access difficult science content by connecting to children's prior knowledge and existing funds of knowledge.

Reading as an Act of Inquiry: Teacher creates opportunities for students to engage with text by reading as an act of questioning text and as one part of a larger scientific inquiry. Reading by actively engaging with the text by questioning, examining claims, and interrogating provided evidence.

Fig. 6.3 Emerging coding categories from data analysis

Table 6.1 Average distribution of language events across the different levels of implementation of the SWH approach

Implementation level	About	Through	Living
Level 1	24.0	20.8	13.2
Level 2	41.1	41.9	30.4
Level 3	60.9	67.1	73.3

Data Analysis Teacher implementation of the SWH approach was scored as level 1 – low, level 2 – medium, or level 3 – high. Next, analyses were conducted for the relevant data sources and presented by SWH implementation level to highlight variation between low, medium and high teacher implementers for the embedded language practices. Previous research indicates that teachers need between 18 to 24 months to be able to successfully implement and sustain the SWH approach (Norton-Meier et al. 2009; Hand 2008; Hand et al. 2015). Consequently, data is presented for those teachers who participated in the project either for 2 or 3 years to give these teachers sufficient time to be comfortable with the approach. As can be seen from Table 6.1 above, there is a distribution of teachers across the three levels of implementation.

Results

The distribution of the teachers with at least 2 years of teaching the SWH approach was 18 teachers at level 2 and 9 teachers at level 3, reflecting previous research that suggests it is difficult for teachers to be consistently at the highest level of implementation.

A critical part of the research was to assess how the teachers were using these particular language events or opportunities as part of the process of embedding language practices within the SWH approach. We believed it was necessary to be able to understand how teachers set up and frame the language environment for the young children as a means of promoting both science concept knowledge and the development of language and multimodal representations. As described above the theoretical framework is centered on *learning about language through using the language as you live the language*. To look more globally at these events, the first layer of analysis was to determine the broader use of each of these major categories: about, through and living. Table 6.1provides the average of the three levels of implementation with respect to these three types of language events: about, through and living.

From Table 6.1, high implementers (Level 3) used all language categories more frequently followed by Level 2 implementers and then Level 1 implementers. In addition to more frequent use, Level 3 teachers also emphasized living the language more than their Level 2 and 1 peers. Specifically, their use of living the language events was 2.41 and 5.5 times as high as their Level 2 and Level 1 peer's use, respectively.

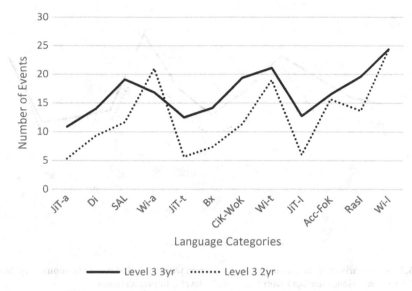

Fig. 6.4 A comparison of the language categories used within the classroom environment between 2 and 3 year experience teachers who were rated a level 3 implementation

In order to examine the role of professional development in teachers' use of the full range of language events, an analysis of the differences between teachers with 2 years of involvement in the project versus those with 3 years of involvement was conducted. Involvement included attendance at the workshops, participation in the school meetings, and implementation of science units that were framed around the SWH approach. The following graphs show the outcome of this analysis. As can be seen at each level of implementation (level 1, 2 or 3), the teachers with 3 years of involvement used more of the embedded language practices required for the SWH approach (Figs. 6.4, 6.5 and 6.6).

Importantly the categories of language use more frequently used by higher implementing teachers included the argumentative categories of dialogical interaction and scientific argument language and the use of more sophisticated words and power verbs. These differences suggest that the focus in classrooms was more on the conceptual knowledge of a topic rather than on science content, a clear shift in seeing reading as an act of inquiry, and an overall increase in the use of writing as a learning tool. Students not only learned about the process of writing but also used writing to help them understand the concepts. Essentially, they were living the language that was an essential element of the environment. This shift in how language is used within these classrooms is particularly interesting because it underscores how early childhood classrooms can become places where sophisticated language skills are developed and used.

While these results highlight the difference between levels and years of experience, we were also interested in how the teachers, regardless of years of experience,

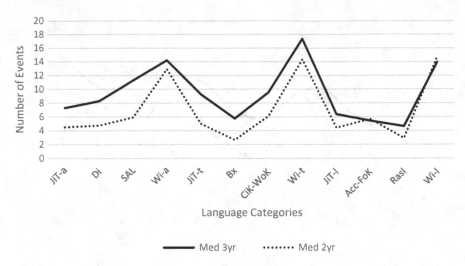

Fig. 6.5 A comparison of the language categories used within the classroom environment between 2 and 3 year experience teachers who were rated a level 2 implementation

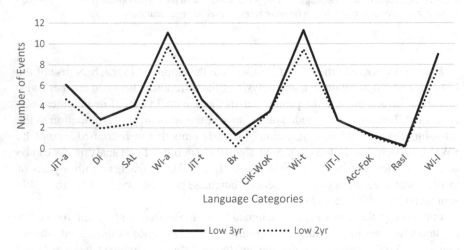

Fig. 6.6 A comparison of the language categories used within the classroom environment between 2 and 3 year experience teachers who were rated a level 1 implementation

at the end of the project were distributed across the levels of implementation. The distribution was Level 1–3 teachers, Level 2–17 teachers, and Level 3–9 teachers indicating that by end of 2 years or more of experience nearly all teachers were at Level 2 or above. This distribution of implementation levels is reflected in the language experiences that the K-3 students are exposed and can be seen in Fig. 6.7.

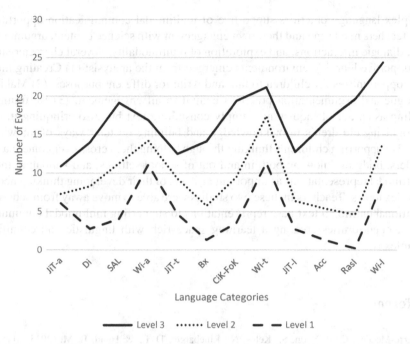

Fig. 6.7 A comparison of different levels of implementation of the SWH approach and the distribution of language categories used

Conclusion

The importance of the Study 2 results frame the Study 1 results and offer an explanation for why young children in SWH classrooms were capable of generating more and different modes of expression and then more cohesively integrating these modes with one another. Student and teacher engagement in a range of different language activities is likely the key to promoting this link between visual and textual representations. While young children are constantly exposed to many different modal forms across books, posters, and other media, the ability to connect these different modes together in a coherent manner is lacking. The demand on the young children to engage in dialogic experiences and link claims to evidence is important – they not only need to learn about this, but they need to live this language. Living the language of argument means that we can generate within students an interest and purpose for reading as an act of inquiry, to use more sophisticated word choices, and deal with "big ideas" as tools for improving their understanding of science. Learning about language through using the language as they live the language promotes the idea that language writ large is an epistemological tool.

The studies also highlight the time that is needed for teachers to adopt these practices in the classroom, ideally 2–3 years of continuous support for their learning. To create a space that validates the ability of young children to engage in the complex language practices supportive of multimodal communication opportunities, teachers need time and their own engagement with science content, argumentation, dialogic interactions, and exploration of multimodality. Several characteristics of supportive learning environments emerged from the analysis: (1) Creating multiple opportunities for children to talk and write for different purposes, (2) Making dialogue and argumentation practices central to all engagement, (3) Re-framing reading as an act of inquiry not simply consuming text but also critiquing it, (4) Recognizing children's prior knowledge and building on their ways of knowing, and (5) Supporting children's thinking through helping them cross over content area borders, bridging knowing both in and out of school settings, and validating their multimodal representations as important artifacts of their developing thinking about complex topics. Teachers in these two studies were able to move away from school's traditional unimodal text base representations to supporting multimodal communication opportunities creating a learning space rich with linguistic and cognitive complexity.

References

Aguirre-Mendez, C. P., Yoon, S., Keles, N., Linebarger, D. L., & Hand, B. M. (2013, March). *Early learners' multiple representation in the context of the science writing heuristic approach.* Paper presented at the annual meeting of the National Association for Research in Science Teaching, Pittsburg, PA.

Akkus, R., Gunel, M., & Hand, B. (2007). Comparing an inquiry based approach known as the Science Writing Heuristic to traditional science teaching practices: Are there differences? *International Journal of Science Education, 29*, 1745–1765.

Bandura, A. (2001). Social cognitive theory of mass communication. *Media Psychology, 3*(3), 265–299.

Barr, R., & Brito, N. (2013). From specificity to flexibility: Developmental changes during infancy. In P. Bauer & R. Fivush (Eds.), *Wiley-Blackwell handbook on the development of children's memory* (pp. 453–479). Chichester: Wiley and Sons.

Bransford, J. D., & Schwartz, D. L. (1999). Rethinking transfer: A simple proposal with multiple implications. In A. Iran-Nejad & P. D. Pearson (Eds.), *Review of research in education* (Vol. 24, pp. 61–100). Washington, DC: American Educational Research Association.

Burke, K., Greenbowe, T., & Hand, B. (2006). Implementing the Science Writing Heuristic in the chemistry laboratory. *Journal of Chemical Education, 83*, 1032–1038.

Conezio, K., & French, L. (2002). Science in the preschool classroom: Capitalizing on children's fascination with the everyday world to foster language and literacy development. *Young Children, 57*(5), 12–18.

Cook, C., Goodman, N. D., & Schulz, L. E. (2011). Where science starts: Spontaneous experiments in preschoolers' exploratory play. *Cognition, 120*, 341–349.

Damon, W., Kuhn, D., & Siegler, R. (1998). *Handbook of child psychology.* New York: Wiley.

Danish, J. A., & Phelps, D. (2011). *The interactional role of kindergarten and first grade students' representational practices.* Paper presented at The Annual Meeting of the American Educational Research Association, New Orleans.

Fantuzzo, J., Sekino, Y., & Cohen, H. L. (2004). An examination of the contributions of interactive peer play to salient classroom competencies for urban Head Start children. *Psychology in the Schools, 41*(3), 323–336.

Ford, M. J., & Forman, E. A. (2006). Chapter 1: Redefining disciplinary learning in classroom contexts. *Review of Research in Education, 30*(1), 1–32.

Gopnik, A. (2012). Scientific thinking in young children: Theoretical advances, empirical research, and policy implications. *Science, 337*(6102), 1623–1627.

Halliday, M. A. K. (1975). *Learning how to mean: Explorations in the development of language.* London: Arnod Press.

Halliday, M. A. K., & Martin, J. R. (1993). *Writing science: Literacy and discursive power.* Pittsburgh: University of Pittsburgh Press.

Hand, B. (Ed.). (2008). *Science inquiry, argument and language: The case for the Science Writing Heuristic (SWH).* Rotterdam: Sense Publishers.

Hand, B., & Keys, C. (1999). Inquiry investigation. *The Science Teacher, 66*(4), 27–29.

Hand, B., Norton-Meier, L., Gunel, M., & Akkus, R. (2015). Aligning teaching and learning: A 3–year study of embedding authentic language and science practices within elementary science classrooms. *International Journal of Science and Mathematics Education*, pp. 1–17. doi:10.1007/s10763-015-9622-9.

Hayne, H. (2006). Bridging the gap: The relation between learning and memory during infancy. In M. H. Johnson & Y. Munakata (Eds.), *Attention and performance XXI: Processes of change in brain and cognitive development* (pp. 209–231). London: Oxford University Press.

Johnson, C. (2011). Hide and seek and the air in the closet: Environments for learning. In B. Hand & L. Norton-Meier (Eds.), *Voices from the classroom: Elementary teachers' experience with argument-based inquiry.* Rotterdam: Sense Publishers.

Kozma, R. (1994). Will media influence learning: Reframing the debate. *Educational Technology Research and Development, 42,* 7–19.

Kress, G. R. (1997). *Before writing: Rethinking the paths to literacy.* New York: Routledge.

Lany, J., & Saffran, J. R. (2013). Statistical learning mechanisms in infancy. In J. L. R. Rubenstein & P. Rakic (Eds.), *Comprehensive developmental neuroscience: Neural circuit development and function in the brain* (Vol. 3, pp. 231–248). Amsterdam: Elsevier.

Lemke, J. (1994). *The missing context in science education: Science.* Paper presented at American Educational Research Association annual meeting, Atlanta, April 1992. Arlington: ERIC Documents Service (ED 363 511), 1994.

Malin, J. (2011). What's the big ideas? Putting concept maps into the hands of your students. In B. Hand & L. Norton-Meier (Eds.), *Voices from the classroom: Elementary teachers' experience with argument-based inquiry.* Rotterdam: Sense Publishers.

Nelson, S. (2011). Science argumentation and the arts. In B. Hand & L. Norton-Meier (Eds.), *Voices from the classroom: Elementary teachers' experience with argument-based inquiry.* Rotterdam: Sense Publishers.

Norris, S. P., & Phillips, L. M. (2003). How literacy in its fundamental sense is central to scientific literacy. *Science Education, 87,* 224–240.

Norton-Meier, L. (2008). Creating border convergence between science and language: A case for the Science Writing Heuristic. In B. Hand (Ed.), *Science inquiry, argument and language: The case for the Science Writing Heuristic (SWH)* (pp. 13–24). Rotterdam: Sense Publishers.

Norton-Meier, L., Hand, B., Cavagnetto, A., Akkus, R., & Gunel, M. (2009). Pedagogy, implementation and professional development for teaching science literacy: How students and teacher know and learn. In M. C. Shelley II, L. D. Yore, & B. Hand (Eds.), *Quality research in literacy and science education: International perspectives and gold standards* (pp. 169–188). Dordrecht: Springer.

O'Keeffe, M. (2010). *Media and the making of scientists.* Unpublished dissertation. University of Pennsylvania, Philadelphia. pp. 1–17. doi:10.1007/s10763-015-9622-9.

Pahl, K., & Rowsell, J. (Eds.). (2006). *Travel notes from the new literacy studies: Instances of practice.* Clevedon: Multilingual Matters Ltd.

Richmond, J., & Nelson, C. A. (2007). Accounting for change in declarative memory: A cognitive neuroscience perspective. *Developmental Review, 27,* 349–373.

Sanders, J. (2011). Implementing science conversations with young learners. In B. Hand & L. Norton-Meier (Eds.), *Voices from the classroom: Elementary teachers' experience with argument-based inquiry.* Rotterdam: Sense Publishers.

Sawada, D., Piburn, M., Judson, E., Turley, J., Falconer, K., Benford, R., & Bloom, I. (2002). Measuring reform practices in science and mathematics classrooms: The reformed teaching observation protocol. *School Science and Mathematics, 102*(6), 245–253.

Sigel, I. E. (1999). Approaches to representation as a psychological construct: A treatise in diversity. In I. E. Sigel (Ed.), *Development of mental representation: Theories and applications* (pp. 3–12). Mahwah: Erlbaum.

Teale, W., & Sulzby, E. (1986). *Emergent literacy: Writing and reading.* Norwood: Ablex.

Weisberg, D. S., Hirsh-Pasek, K., & Golinkoff, R. M. (2013). Guided play: Where curricular goals meet a playful pedagogy. *Mind, Brain, and Education, 7*(2), 104–112.

Yelland, N. J., O'Rourke, M. E., Lee, L., & Harrison, C. (2008). *Rethinking learning in early childhood education.* Buckingham: OUP.

Yoon, S. (2012). *Dual processing and discourse space: Exploring fifth grade students' language, reasoning, and understanding through writing.* Unpublished dissertation, University of Iowa, Iowa City.

Chapter 7
Examining the Impact of Multimodal Representation Instruction on Students' Learning of Science

Jeonghee Nam and Hyesook Cho

Introduction

There is growing recognition in science education that student learning of science is directly related to the learning of science language. Learning science involves not only understanding science concepts, but also the ability to communicate scientific concepts and ideas (Yore et al. 2003). Language can help students develop and construct meaning about science concepts as well as provide the vehicle for students to communicate about the scientific understanding they have developed. Building on the idea of the importance of language as a way to improve students' understanding of science, recognition has been given to the important role multimodal representations play in science.

Several studies provide support for the claim that students' writing activities have a positive impact on student learning (Hand et al. 2001, 2007; Rivard 1994). Writing is a powerful and an epistemological tool for learning science through language. Effective writing both displays and provides an avenue for the construction of an understanding of science concepts. Using multimodal representation as a part of this writing may play a role in supporting reasoning and thinking.

One way to extend this notion of writing-to-learning in science currently being explored is to encourage students to use multiple modes of representation. As pointed out by Emig (1977), enactive writing as the reformulation of ideas involves the representation of ideas in images as well as in verbal symbols. Lemke (1998, p. 87) further suggested that "to do science, to talk science, to read and write science, it is necessary to juggle and combine in canonical ways verbal discourse, mathematical expression, graphical-visual representations, and motor operations in the natural world." Scientists, in particular, rely heavily on graphical representations

J. Nam (✉) • H. Cho
Department of Chemistry Education, Pusan National University, Busan, Korea
e-mail: jhnam@pusan.ac.kr; rabeey@naver.com

© Springer International Publishing Switzerland 2016　　　　　　　　　　　　117
B. Hand et al. (eds.), *Using Multimodal Representations to Support Learning in the Science Classroom*, DOI 10.1007/978-3-319-16450-2_7

of objects of interest, e.g., photographs, drawings, maps, graphs, and tables (Roth and McGinn 1998). To appropriately and effectively construct, mobilize, and combine various forms of scientific representations also raises the degree of persuasiveness of presented arguments (Latour 1990). Kozma et al. (2000) found that chemists move across and use different representations to understand the chemical phenomena of their investigations and to argue for, explain, and justify their findings. A correspondence between features of different representations can serve as a warrant for confirming or disconfirming conclusions about findings. They suggest that students should use multiple representations to explain their findings and argue for their conclusions.

Research in this area of multimodal representations as a part of writing-to-learn activities has centered on student use of different modes of representing information such as diagrams, pictures, math equations, tables, graphs. Students need to learn how to select appropriate representations for addressing particular needs, and be able to judge their effectiveness in achieving particular purposes. Thus, the use of multimodal representation becomes important because the language of science is multimodal, that is, concepts of science are described by and across different modal representations (Yore and Treagust 2006). The intent of this study was to link the two previously mentioned research areas: the use of multimodal representations and writing-to-learn. In framing the discussions on multimodal representation, Lemke (1998) has stated "Science is not done, is not communicated, through verbal language alone. It cannot be." He adds that scientists "combine, interconnect, and integrate verbal text with mathematical expressions, quantitative graphs, information tables, abstract diagrams, maps, drawings, photographs, and a host of unique specialized visual genres seen nowhere else"(p. 89).

The use of multimodal representations could play an important role in helping students construct a deeper understanding of science concepts (Nakhleh and Postek 2008). However, past research studies have focused on the impact of multimodal representations which were provided or introduced by instructors or through technology-based instruments on student understanding of scientific concepts (Kozma and Russell 2007). Very few studies have focused on representations created by the students themselves.

As there is growing agreement about the role of language in science education, writing activities in science teaching has been also recognized and utilized to improve student understanding of science concepts. With this focus on language in teaching in science education, embedding multimodal representations has become more recognized as a learning tool to promote student understanding science concepts. However, there has been little research attention focused on using writing-to-learn strategies with embedded multimodal representations as a means to help improve students understanding of science in actual classroom setting. Prain (2006) argues for the necessity of examining the pedagogical practices that help shape the writing activities for students. Airey and Linder (2006) suggested that learning science is more than just exposure to the modes of science; rather it should be about allowing students opportunities in "using the disciplinary discourse to make meaning for themselves", that is, using the constellation of modes for a topic to construct understanding of the topic. These statements support the implementation of teaching practices in which students are asked to communicate multimodally in science.

Science as a discipline is multimodal. That is, it involves the negotiation and production of meanings in and through different modes of representation (Murcia 2010). New scientific understanding is generated through multiple representations of ideas, affective responses and evidenced based judgments (Tytler 2007). Using multimodal representations, constructing multiple modes and converting between multiple representations are fundamental to successful learning (Spiro and Jehng 1990). Student writing has also been established to be a useful way of learning science, and while students are engaged in writing activities, they may also mathematical equations, chemical equations, graph, table, and drawings (Airey and Linder 2009). The use of these multimodal representations would be likely be related to students' deeper understanding. Thus, students need to understand modal diversity in representations of science concepts and processes, be able to translate between different modes, as well as understand their coordinated use in representing scientific knowledge (Waldrip et al. 2006). Therefore, the science classroom should include opportunities for students to use science language including the different types of representations typically employed by scientists as they communicate. More focus is needed on how we can engage students in constructing understanding by requiring them to link the various modal representations of science and integrating meanings across modes.

With respect to this greater need for requiring students to utilize multiple modes of representation in their scientific communication, we attempted to develop and implement instruction aimed at helping students better embed multimodal representations in their science writing as a pedagogical strategy to help improve students understanding of science.

In this study, students in the treatment group received specific instruction encouraging the use of multimodal representations in their science writing. The instruction was designed to encourage students to not only use multiple modes in their science writing, but also employ strategies to effectively link the multiple modes together. The purpose of this study was to investigate the impact of this instruction on student writing as well as on student conceptual understanding. We believe that if the process of embedding multiple modes in written text was beneficial in terms of student's use of multimodal representation and understanding of scientific concept, then we would have an process whereby we can encourage Korean teachers to have their students use multiple representations more in their learning. If the results were positive then we should encourage teachers to utilize this strategy as one way to promote the effective use of multimodal representations in the classroom. Research questions that guide this study are as follows:

1. What are the impacts of multimodal representation instruction on the effective use of multimodal representations in students' writing?
2. What are the impacts of multimodal representation instruction on students' conceptual understanding?

Methods

Participants

This study employed a quasi-experimental design with treatment and control groups from a middle school. The middle school, characterized by a low socioeconomic level, was located in a metropolitan city in Korea, and received low scores in the Korean national standardized achievement test. Participants of this study were seventy- seven 8th grade students who were members of three pre-existing classes at the middle school taught by one of researchers. Students from two classes (50 students) were assigned to the treatment group. Treatment students received three steps of instruction encouraging the using of multimodal representations. One class (27 students) was assigned to the control group and did not receive the instruction. The normal instructional approach to teaching of science was provided throughout the treatment period for both the control group treatment groups except when the treatment group received the instruction on multimodal representations. This study was conducted during the second semester of the 8th grade science curriculum. The teacher participating in this study had 12 years of teaching experience at the secondary level and was enrolled in doctoral degree course at the time of the study. She had been involved in a multimodal representations research project for 3 years at the time of this study.

Multimodal Representation Instruction

The instruction designed to encourage the use of multimodal representations involved three steps of instructional process. The first step of the instruction on the use of multimodal representations was a recognition process. Students began this step by identifying different types of modes used in textbooks. Initially they were asked to recognize the difference between a single-mode representation of a specific concept that used text only and multimodal representations, in which pictures tables, charts, diagrams, graphs or formula were utilized with text to describe a concept. During this recognition step, while students were reading the samples of scientific communication, they were encouraged to identify the characteristics of each mode and differences between single-mode representations and multimodal representations. Students were then asked which sample helps them understand the scientific concept or idea better and why the sample they choose lead to better understanding. Throughout this step, students recognized how information in a scientific source such as a textbook or science-related book is best represented and explained. Students were also asked to both identify distinguishing features between each mode and to identify the key differences between the individual modes.

The second step in the process of instruction was related to using multiple representations in an interpretative process whereby students learned to convert one mode to other modes through understanding the characteristics of each mode and the purposes of specific types of modes. Emphasis was placed on encouraging the students to relate the different modal forms to each other, that is, to examine hoe the modes used are related. Through this process students came to understand what they should do to help people understand how modes help represent the ideas discussed in text. In this step of the instructional process, students were asked to seek out the various representations used in science textbooks and consider how the multimodal representations were used to construct and justify evidence-based claims. The students were also asked to analyze how the multimodal representations were used in textbooks in terms of constructing and justifying the big idea of the unit. In this step, students identified types of representations used in the textbook and how the representations help them understand the concept presented as a way of beginning to evaluate effective multimodal use.

The third step of instruction was an application process using multimodal representations. In this application process, students had an opportunity to apply their understanding of multimodal representations to their own writing to communicate their understanding of a scientific concept. During this third step, students were asked to draw a concept map of the unit using multimodal representations. Students completed their own concept map and had group discussions about the role and importance of multimodal representational information in writing. This activity was intended to help students recall important science concepts from the just completed unit before writing a summary and consider how to represent these concepts using a range of modes such as graphs, tables, diagrams, equations, and photographs. The teacher encouraged students to use multiple modes for presenting the concept(s) in the concept map to gain practice representing scientific ideas with different modes. Concept maps were used as an organizer to illustrate students' knowledge and therefore provided an appropriate tool to both utilize multiple representations and to begin to summarize ideas about the unit of study and the targeted concepts before the summary writing experience. Students were then asked to write a summary of the unit based on the concept map they drew. The teacher asked students to explain to their friends what they understood about the unit but did not require students to use one mode other than text in their summary writing. Students were allowed to spend one class period to complete their writing (Table 7.1).

Students in the treatment group also participated in a small group discussion to reflect on their understanding of multimodal representations and how they help students understand science concepts. During a series of discussions, students called on their experience in analyzing writing samples given by the instructor to facilitate the discussion. During this activity, the teacher provided an article from a science magazine to the students and asked them to investigate how information in the article was presented to readers. The intent was to encourage the students to discuss how multimodal representations were used and whether these modal representations were useful in helping them understand the ideas presented in the article. Then students in the small groups then shared their ideas with the entire class. The small group discussion activity took place within a single one hour lesson.

Table 7.1 Summary of the three steps involved in the process of multimodal instruction

Step	Instructional goal	Student activity
Recognition	Identify different types of modes Recognize the difference between a single mode representation and multimodal representations	Read a reading materials about density identify individually different types of modes and the characteristics of each mode and differences in the reading materials
Interpretation	Understand characteristics of each mode and the purposes of specific types of modes Recognize the importance of using multimodal representations for evidence in support of a claim	Examine in pairs the textbook section on solubility individually and answer the questions provided by a teacher And convert one mode to other mode Examine the textbook section on properties of matter unit and answer the questions provided by a teacher
Application	Understand the role of multimodal representations and use multimodal representations in writing	Draw a concept map individually at the end of stimulus and response unit Write a summary individually at the end of electricity unit Read newspaper articles and respond in the group to the questions provided by a teacher

The multimodal representation instruction was implemented during the second semester of the school year. In the second semester, four units were taught including 'properties of matter', 'stimulus and response', 'electricity', and 'mixtures' that were all based on the national science curriculum sequence. The unit of 'properties of matter' contains concepts related to identifying substances and characterizing materials by melting point, boiling point, density, and solubility. The 'stimulus and response' unit includes the structure and function of the sensory organs, the structure and function of the nervous system and neuron, and the understanding of reaction path for the stimulation. The 'electricity' unit includes concepts related to understanding of current and charge conservation, concepts dealing with materials that are either a conductor or an insulator, as well as the concept of resistance, and Ohm's law. The 'mixture' unit includes the concept and characteristics of mixtures and compounds, and separation of mixtures.

During the first lesson of the 'properties of matter' unit, students were given reading material from a science textbook and resource book provided by the teacher. The reading material was centred on the concept of density which the students had not previously engaged with and included multi modal examples such as a table, an equation, and pictures with text. The students were asked to identify the types of modes in the reading materials as part of the recognition process. They were then asked to compare each mode and find the differences between them. Figure 7.1 shows the content of the reading material selected from a science textbook provided by the teacher.

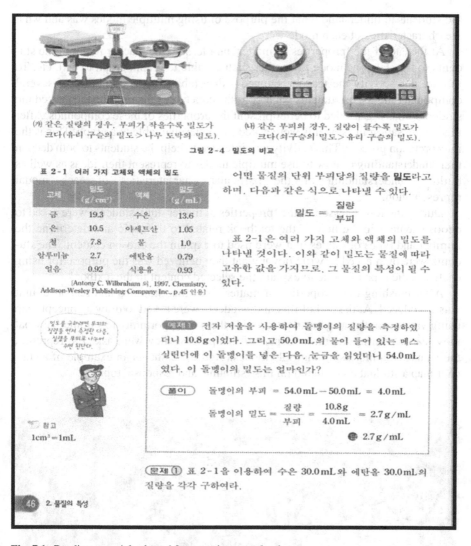

Fig. 7.1 Reading material selected from a science textbook

At the middle of the 'properties of matter' unit when the solubility concept was introduced to the students, students were asked to examine the textbook section on solubility and find different modes used to describe solubility and analyze how the modes were used. They were then requested to figure out the difference in terms of clarity and ease of understanding when one mode was used such as text, table, graph, figure, and diagram and when more than two modes were used. The teacher asked the following questions to the students; 'what is easier to understand text when presented with a graph or figure, or only text?, 'Why?', and 'Which one is better to understand about the solubility? During this activity, the teacher assisted the students to understand what the purpose of using multiple modes was and what the characteristics of each mode were.

At the end of the 'properties of matter' unit lesson, the teacher required the students to translate one mode to another such as table to graph, graph to text, etc. To initiate this activity the teacher showed a data table for the solubility of several compounds and asked students what does the data table tells us? She then asked the students what was needed to better present the properties of these compounds. After this conversation the students were asked to convert the table to a graph. This is the interpretation process. This activity was intended to help the students to both deepen their understanding of how to use multiple modes to represent their ideas as well as building understanding of the particular characteristics of each modal representation.

During the fourth lesson of the 'properties of matter' unit, students were asked to choose content in the unit of the textbook related to the topic and determine the number and types of modes that were used to explain the chosen content. The students were then asked to write down what they believed was the purpose for using each specific type of mode to explain the selected content in the textbook.

After finishing the 'properties of matter' unit, the 'stimulus and response' unit was introduced. At the end of this unit, students were asked to draw a concept map using various modes as a means to provide them an opportunity to summarize what they learned during the unit. As part of this process they were asked to represent their understanding using multiple modes. Figure 7.2 shows an example of a concept map a student constructed on this 'stimulus and response' topic.

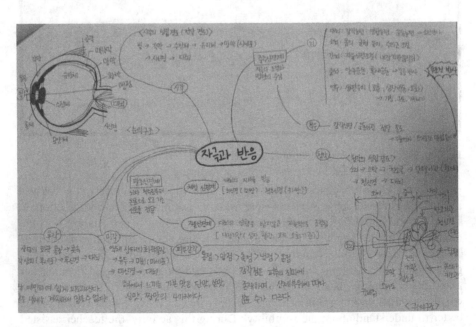

Fig. 7.2 Student's concept map related to the topic of stimulus and response

At the end of the 'electricity' unit, the teacher gave a writing task to the students to write a summary of the unit to explain big idea of the unit to a friend who has not studied the topic of 'electricity'.

When all four units including the 'mixtures' unit ended, the teacher provided a science-related article taken from a Korean newspaper as well as supplementary materials to students. The teacher also provided them with background scientific information and a related data table followed by a prompt. The prompt presented a claim by an expert who provided an explanation that included information from the data provided. The teacher then asked several questions such as "what was the reporter's claim?", "what did the reporter use to persuade the readers?", and "what is more persuasive?". Then the teacher led students in a group discussion and asked them to make a presentation to the class about the claims and evidence presented in the article.

In the control group class, the concept map activity was introduced by the teacher at the end of the 'stimuli and response' unit. For the electricity unit the teacher was the person who summarized the unit by asking series of questions confirming whether students understood the science concepts in the unit (Table 7.2).

Data Collection and Analysis

Multimodal Representation Assessment All students in both the treatment and control groups were assessed for multimodal representational use before and after the study through a pre-test and post-test design. Each student from both the multimodal representations instruction treatment group and the control group was assigned a writing task regarding 'states of matter' (for the pretest) and 'electricity' (for the posttest). In the writing task, students were asked to explain to their friends what they understood about the topic given. Students completed this writing task in a single 45-min class period at the beginning of the semester (pre-test) and then at the end of the semester (post-test).

The writing task was scored using an analysis framework for multimodal representations. This analytical framework was developed for this study based on a multimodal writing task assessment rubric utilized in previous research projects (McDermott 2009; Gunnel et al. 2009). The analytical framework focused on four aspects of the written product: an assessment of the text, an assessment of the cohesiveness between the alternative modes and the text, an assessment of the use of multimodal representations outside of text, and an assessment of the degree of embeddedness (how well the modes are integrated) of multimodal representations within the text. The summary writing for the writing task that was assessed involved students communicating their conceptual understanding of the topic. The intent of the summary writing was to provide an avenue for students to display their understanding and to use the framework developed through engagement in the unit activities. These activities had included the use of multimodal representations in the treatment group, but did not include their use for the control group.

Table 7.2 Summary of the science units involved with the multimodal student activity involved

Step	Unit	Major concept	Student activity	Lesson	Time period	Position in unit
Recognition	Properties of matter	Identifying substances	Read a reading materials about density and identify individually different types of modes and the characteristics of each mode and differences in the reading materials	One	20 min	Beginning
Interpretation	Properties of matter	Characterizing materials by properties of matter / Melting point, boiling point, density, solubility	Examine in pairs the textbook section on solubility individually and answer the questions provided by a teacher and convert one mode to other mode	One	45 min	Middle
			Examine the textbook section on properties of matter unit and answer the questions provided by a teacher	One	25 min	End
Application	Stimulus and response	Structure and function of the sensory organs / Structure and function of the nervous system and neuron / Reaction path for the stimulation	Draw a concept map	One	45 min	End
	Electricity	Current, charge, conservation / Conductor and insulator / Resistance	Write a summary	One	45 min	End
	Mixtures	Ohm's law / Mixtures / Compounds / Separation of mixtures	Read newspaper articles and respond in the group to the questions provided by a teacher / Presentation	One	45 min	End

The text assessment portion of the analysis examined student use of text in explaining ideas about the topic of interest. This section assessed seven specific components of the written product which included grammar, coverage of required topics, the number of big ideas linked to the topic, accuracy of the described science concepts, completeness of meaning, logical order of the text, utilization of appropriate language for the audience utilized in the text, and identification of key terms. Each component was scored on a 0–2 scale based on the evidence of use, except the number of big ideas scoring category which was scored by the number of big idea which students provided. Students were given zero points for any category in which no evidence of that characteristic was present.

The assessment of cohesiveness examined student use of alternative modes and their link to the text. This section assessed four specific components which included text tied to alternative modes used, alternative modes linked to each other, examples carried throughout the writing sample, and alternative modes linked to more than one topic. We were concerned that the students attempt to ensure that the main conceptual idea was continually addressed throughout the writing sample. Each component was scored on a 0–2 scale based on the evidence of use. Students were given zero points for any category in which no evidence of that characteristic was present.

The final two sections in the assessment rubric focused on assessment of the non-text aspects of the written products. The section assessing use of multimodal representations examined how alternative modes outside of text, such as diagrams, graphs, equations, figures, and tables, were used to explain an idea. Each mode was scored based on the frequency. One point was awarded on the assessment for each mode outside of text. Students were given one point for each mode that was utilized that was not text. The section assessing degree of embeddedness of multimodal representations examined how students linked their modes to text. This section assessed seven specific components which included (1) the presence or absence of a caption, (2) the placement of modes outside of text next to appropriate text, (3) the reference to modes outside of text in the text, (4) the originality of modes outside of text in text, (5) the scientific accuracy of the information of the modes outside of text, (6) necessity of modes for explanation, and (7) the conceptual connection of the mode outside of test to the information in the text. Each component was scored on a 0–2 points scale in which 0 points were awarded for no attempt to include the characteristic, 1 point was awarded for a limited attempt to include the characteristic, and 2 points were awarded for complete inclusion of the characteristic.

To ensure consistency on the rating of the students work, the instructor who was the person responsible for leading the multimodal representation instruction for the students, along with two doctoral students randomly chose 20 writing task samples from the post-test and discussed each component of the analysis framework until consensus on the scores was achieved. Then, the instructor then scored all the writing task samples from pre-test and post-test.

Science Content Assessment All students in both groups participating in this study were also assessed in regard to their science conceptual understanding through a series of two Science Concept Tests developed by the researchers. Students were assessed with these instruments before and after the study for pre-test and post-test conceptual understanding scores. Each test was composed of 10 two-tiered items which required the student to answer a multiple-choice question and then justify their response by explaining why they selected their particular response. Test items for the pre-test were developed by the researchers based on the 7th grade science concepts and in consultation with doctoral students who taught at the middle school level. Post-test items were also created by the researchers based on the concepts covered in the unit 'electricity' and through consultation with the same doctoral students. Each item was scored on a 0–3 scale according to the student response. A complete and correct response with the correct choice was worth 3 points. A partially correct response with the correct choice was worth 2 points. A correct choice with no response or wrong response was worth 1 point. Students were given 0 points if they provided the correct choice without any response.

The same instructor who led the multimodal representations encouraging activities scored all the content tests. Inter-rater reliability of scoring was calculated by the teacher rescoring a randomly selected group of 20 different responses from the student tests and comparing the result of the first scoring with that of the second. The scores for each question were compared to determine inter-rater reliability. An average of agreements for all items for the pre-test and post-test was 0.9.

Results

Multimodal Representation Scores of the Treatment and the Control Groups

To investigate differences in multimodal representation in the writing tasks between the two groups, a one way ANCOVA was conducted on post-test scores of writing task with pre-test scores used as a covariate. Each category of text assessment, the use of multimodal representations, and embeddedness of multimodal representations, as well as a total score for the writing task (the combined total of the previously mentioned three sections) were used as dependent variables. Treatment condition served as the independent variable. Table 7.3 provides all the mean scores and standard deviations for the assessment of the writing tasks. The results of this analysis indicate that the treatment group students (mean=33.64, SD=11.73) scored significantly higher than the students in the control group (mean=27.70, SD=13.11) on the total writing task score (F (1, 75)=7.34, $p<0.05$). The results also indicated that there was significant difference in the scores of the text assessment between the groups with the treatment group (mean=12.82, SD=12.83) outperforming the control group (mean=10.85, SD=4.79) (F (1, 75)=5.60, $p<0.05$).

There was also a significant difference in the scores related to use of multimodal representations between the groups with the treatment group (mean=8.98,

Table 7.3 Scores on the writing task for pre- and post-test

	Pre-test			Post-test		
	n	Mean	S.D.	n	Mean	S.D.
Text assessment						
Treatment	50	9.98	2.52	50	12.82	12.83
Control	27	10.33	2.91	27	10.85	4.79
Use of multimodal representations						
Treatment	50	3.62	2.72	50	8.98	4.13
Control	27	3.15	2.60	27	7.97	3.87
Cohesiveness						
Treatment	50	3.40	1.12	50	5.34	2.79
Control	27	3.52	1.05	27	4.15	2.92
Embeddedness of multimodal representations						
Treatment	50	4.42	2.22	50	6.50	2.41
Control	27	4.93	2.40	27	5.62	2.88
Total scores of writing task						
Treatment	50	21.42	6.99	50	33.64	11.73
Control	27	21.93	7.26	27	27.70	13.11

SD $= 4.13$) outscoring the control group (mean $= 7.97$, SD $= 3.87$) (F $(1, 75) = 3.28$, $p < 0.1$).

There was a significant difference detected between the treatment group (Mean $= 5.34$, SD $= 2.79$) and control group (Mean $= 4.15$, SD $= 2.92$) (F $(1, 75) = 4.79$, $p < 0.05$) in the scores of the cohesiveness. There was also a significant difference detected between the treatment group (Mean $= 6.50$, SD $= 2.41$) and control group (Mean $= 5.62$, SD $= 2.88$) (F $(1, 75) = 4.17$, $p < 0.05$), in the scores of the embeddedness of multimodal representations.

Figure 7.3 provides an example of the difference between students' writing from the treatment group and the control group for the post-test. The student from the treatment group used a table, a drawing, a graph, and an equation with text to explain the fundamental concepts associated with the topic requested in the writing task. The student from the treatment group tried to link the table with the text and then expanded on the explanation with a drawing. The student also added an equation based on the explanation provided already. This means the student from the treatment tried to integrate each of the modes used in the unit in order to provide a rich explanation of the concept of electricity as requested in the writing task. On the other hand the students from the control group provided equations and drawings without text. There was very little intention to link each mode. The writing samples also show the difference of the cohesiveness between the students of the two groups. The student from the treatment group presented a table to compare the characteristics of a positive charge and negative charge, a figure to show process of electrostatic induction phenomenon and a formula to present Ohm's law. This means the student tried to link many alternative modes used with text and attempted to connect

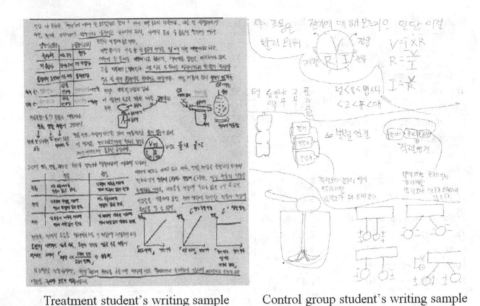

Treatment student's writing sample   Control group student's writing sample

Fig. 7.3 Students writing samples from treatment and control groups

the table and graphs to explain the circuit suggested charge electrify triboelectric, electrostatic induction by giving an example of the interaction balloon and metal rod. The student used a table in order to illustrate the concept of the series circuit and the parallel circuit and presented a graph to show the relationship between current and voltage in accordance with the number of batteries used. The student used various modes and continually addressed the main concept of electricity from charge to circuit throughout writing. The examples shown below represent students who scored at equivalent levels on a pre-Science Concept test. For Fig. 7.3 these students were in the medium level (on three level performance: low, medium, high), while for Fig. 7.4 the students were in the high level.

The following examples show the differences between students' writing from the treatment and the control group in terms of each component in embeddedness of multimodal representations from the analytical framework for multimodal representations of writing task.

Figure 7.4 shows the differences of degree of embeddedness of multimodal representation between the treatment and the control groups. The student from the treatment group presented a caption for the graph located next to text and referred to in text as 'I will show the picture to make it easier to understand'. All types of modes the student presented were accurate and conceptually connected to each of the other modes.

Overall, the results of this analysis indicate that students who were involved in the multimodal representation instruction were significantly better in terms of the four aspects which are test, cohesiveness, use of multimodal representations and

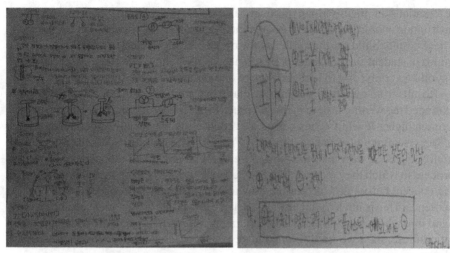

Treatment student's writing sample   Control group student's writing sample

Fig. 7.4 Students writing samples about the degree of embeddedness of multimodal representation

embeddedness of multimodal representations of writing as well as obtaining better scores on the measure of science conceptual understanding.

Conclusions and Implications

The results of this study support the contention that learning to use multiple modes for representing science information is beneficial for student's conceptual understanding, and therefore, students should be encouraged to embed multiple modes of representation in science writing. The activity associated with effectively embedding multiple modes of representation in text likely encourages a process in which students must translate the information dealt with in class into an appropriate "language" for their own understanding, then again into an appropriate "language" to display this understanding in a multimodal representation through writing task to different audiences.

Results of this study also indicate that students in the treatment group were better at utilizing multimodal representations in their written products. Using the multimodal representation framework effectively led the students to integrate various modes in their product and to appropriately represent science information with the various modes. This ability to both use these modes and link them together appeared to be linked to better scores for conceptual understanding. We would suggest that involvement in these multimodal tasks does indicate that the students are constructing a richer and stronger scientific understanding. If, as this study indicates, the presence of instruction encouraging students to effectively embed multimodal representations in their written products facilitates this process, then this type of instruction would represent a pedagogical approach that should be encouraged. These results build on the earlier work on the use of writing to learn strategies to improve student understandings of science, as they show that there is a need to go beyond text only, and to include a range of multimodal representations.

Results from our study support the claim that students in the treatment condition extended the writing task to include a much broader representational emphasis with a much higher level of cohesiveness and connection between the alternative modes of representing information through writing. When students effectively embed multiple modes with text, they are more likely to engage in a beneficial cognitive process in which they more deeply and accurately assess their own understanding of the concept before they determine how to best represent this concept to an outside audience. This is not a trivial task. We believe that if this engagement can be encouraged through instruction aimed at encouraging more effective multimodal representational use, then we need to encourage pedagogy associated with this in science classrooms.

References

Airey, J., & Linder, C. (2006). Language and the experience of learning university physics in Sweden. *European Journal of Physics, 27*(3), 553–560.

Airey, J., & Linder, C. (2009). A disciplinary discourse perspective on university science learning: Achieving fluency in a critical constellation of modes. *Journal of Research in Science Teaching, 46*, 27–49. doi:10.1002/tea.20265.

Emig, J. (1977). Writing as a mode of learning. *College Composition and Communication, 28*, 122–128.

Gunel, M., Hand, B., & McDermott, M. (2009). Writing for different audiences: Effects on high school students' conceptual understanding of biology. *Learning and Instruction, 19*(4), 354–367.

Hand, B., Prain, V., & Yore, L. (2001). Sequential writing tasks' influence on science learning. *Studies in Writing, 7*, 105–129.

Hand, B., Hohenshell, L., & Prain, V. (2007). Examining the effect of multiple writing tasks on Year 10 biology students' understandings of cell and molecular biology concepts. *Instructional Science, 35*(4), 343–373.

Kozma, R., & Russell, J. (2007). Students becoming chemists: Developing representational competence. In J. Gilbert (Ed.), *Visualization in science education*. Netherlands: Springer.

Kozma, R. B., Chin, E., Russell, J., & Marx, N. (2000). The role of representations and tools in the chemistry laboratory and their implications for chemistry learning. *Journal of the Learning Sciences, 9*(2), 105–143.

Latour, B. (1990). Drawing things together. In M. Lynch & S. Woolgar (Eds.), *Representation in scientific practice* (pp. 19–68). Cambridge, MA: MIT Press.

Lemke, J. (1998). Multiplying meaning: Visual and verbal semiotics in scientific text. In J. Martin & R. Veel (Eds.), *Reading science: Critical and functional perspectives on discourses of science* (pp. 87–113). London: Routledge.

McDermott, M. A. (2009). The impact of embedding multiple modes of representation on student construction of chemistry knowledge. Ph. D. Dissertation, University of Iowa, Iowa, USA.

Murcia, K. (2010). Multi-modal representations in primary science: What's offered by interactive whiteboard technology. *Teaching Science, 56*(1), 23–29.

Nakhleh, M. B., & Postek, B. (2008). Learning chemistry using multiple external representations. In J. K. Gilbert, M. Reiner, & M. Nakhleh (Eds.), *Visualization: Theory and practice in science education*. Netherlands: Springer.

Prain, V. (2006). Learning from writing in secondary science: Some theoretical and practical implications. *International Journal of Science Education, 28*(2–3), 179–201.

Rivard, L. P. (1994). A review of writing of learning to learn in science: Implication for practice and research. *Journal of Research in Science Teaching, 31*(9), 969–983.

Roth, W. M., & McGinn, M. K. (1998). Inscription: Toward a theory of representing as a social practice. *Review of Education Research, 68*, 35–59.

Spiro, R., & Jehng, J. (1990). Cognitive flexibility and hypertext: Theory and technology for the nonlinear and multidimensional traversal of complex subject matter. In D. Nix, & R. Spiro (Eds.), *Cognition, education and multimedia: Exploring ideas in high technology* (pp. 163–205). Hillsdale: Lawrence Erlbaum Associates.

Tytler, R. (2007). School innovation in science: A model for supporting school and teacher development. *Research in Science Education, 37*, 189–216.

Waldrip, B., Prain, V., & Carolan, J. (2006). Learning junior secondary science through multimodal representations. *Electronic Journal of Science Education, 11*(1), 87–107.

Yore, L. D., & Treagust, D. F. (2006). Current realities and future possibilities: Language and science literacy—Empowering research and informing instruction. *International Journal of Science Education, 28*, 291–314.

Yore, L. D., Bisanz, G. L., & Hand, B. M. (2003). Examining the literacy component of science literacy: 25 years of language arts and science research. *International Journal of Science Education, 25*(6), 689–725.

Chapter 8
Developing Multimodal Communication Competencies: A Case of Disciplinary Literacy Focus in Singapore

Kok-Sing (Kenneth) Tang, Caroline Ho, and Gde Buana Sandila Putra

Introduction

In science education, there is a growing understanding that learning science involves developing a repertoire of disciplinary-specific literacy skills to engage with the knowledge and practices of the scientific community (Kelly 2008). Such 'disciplinary literacy', or the specific ways of talking, reading, writing, doing, and thinking valued and used by the discipline (McConachie et al. 2006; Moje 2007), is central rather than peripheral to the development of scientific understanding (Norris and Phillips 2003). For decades, researchers from multiple disciplines have shed light on the language and discursive features of academic science (Halliday and Martin 1993; Lemke 1990) as well as pioneering various reading and writing strategies to help students master scientific discourse (Hand et al. 1999; Yore and Shymansky 1985). However, in more recent years, there has been increasing attention toward the role of visual, graphical, mathematical, and gestural modes of representation in scientific communication (Kress et al. 2001; Lemke 1998). Research in this area reveals how each mode of representation plays a unique function in representing different aspects of scientific meaning. More studies are also beginning to show how scientific knowledge in specific content consists of a characteristic and recognizable pattern of relationships among multimodal representations (e.g., Hubber et al. 2010; Tang 2011; Tytler et al. 2006).

Building on our increasing understanding of the role of language and representations in science, current research has begun to focus on developing students' competencies in disciplinary literacy. With the aim of raising the literacy levels of

K.-S. (Kenneth) Tang (✉) • G.B.S. Putra
National Institute of Education, Nanyang Technological University, Singapore, Singapore
e-mail: koksing.tang@nie.edu.sg

C. Ho
English Language Institute of Singapore, Ministry of Education, Singapore, Singapore

© Springer International Publishing Switzerland 2016   135
B. Hand et al. (eds.), *Using Multimodal Representations to Support Learning in the Science Classroom*, DOI 10.1007/978-3-319-16450-2_8

students in all subject areas, there is now a growing recognition of the importance of disciplinary literacy in several national curricula in the learning of all subject areas (e.g., US Common Core Standards, New South Wales National Curriculum). For instance, the Common Core Standards in the United States underscore the importance of literacy in preparation for college and life. In its 'Standards' document for English Language Arts (ELA) and Literacy in History/Social Studies, Science and Technical Subjects, teachers of these subjects are expected to use their "content area expertise to help students meet the particular challenges of reading, writing, speaking, listening, and language in their respective fields" (Council of Chief State School Officers 2010, p. 3).

The purpose of this chapter is to report on findings from a recent research study that focuses on the development of disciplinary literacy in Singapore. As part of a national curriculum shift toward subject-specific communication skills, this research study aims to help science teachers' focus on the teaching of disciplinary language in their classroom. In particular, we worked with teachers to develop pedagogical activities and strategies to help grade 9 (secondary 3) students in physics and chemistry write scientific descriptions and explanations based on observable phenomena. The activities required the students to interpret, translate, and integrate multimodal forms of representations that were introduced at various stages of the science lesson. Through the illustration of selected lesson enactments, we intend to discuss the role of multimodal activities in science disciplinary literacy teaching, as well as highlight the pedagogical issues in implementing disciplinary literacy within the science classrooms.

Theoretical Perspectives

Our research is informed in part by the theory of social semiotics (Lemke 1990), which posits that language and all other symbol systems (e.g., images, gestures) function as meaning-making resources for people to make different kinds of meaning in any social context. Social semiotics is a generalized theory that derives from earlier work in systemic functional linguistics (SFL; Halliday 1978). SFL has been widely applied in science education to investigate the nature of language in science classroom discourse. Early work in the 1990s examined the relationship between a text's linguistic function and students' content development in science. For instance, Lemke (1990, p. 12) identified the repeated and characteristic "thematic pattern of semantic relationships" of words and utterances that constitutes what one would recognize as talking about a particular topic in science. The peculiar features of the specialized language of science were also elaborated by Halliday and Martin (1993), who shed light on several unique linguistic features of scientific texts that present a challenge to students' learning of science. These features include interlocking definitions, technical taxonomies, lexical density, and nominalization. Other researchers (e.g., Schleppegrell 2004; Veel 1997) provided rich descriptions of science genres, such as report, exposition, explanation, and experimental procedure, which students typically go through and need to learn in science lessons. Within the genre of explanation, Unsworth (2001) further examined how different language choices optimize the effectiveness of science explanations in school texts. According to Moje (2007), the SFL approach to disciplinary literacy focuses on teaching the linguistic processes of the disciplines. This involves guiding students through the process of highlighting the grammatical and lexical features of texts, deconstructing various text genres, and jointly constructing new texts using the features of the disciplinary language.

In more recent years, multimodality – the study of multiple modes of representation- has further expanded research from SFL to incorporate other semiotic resource systems such as images, graphs, symbols and gestures (Jewitt 2008). This area of work brings to attention the importance of multimodal integration in science classroom teaching and learning. For instance, Kress et al. (2001) documented the complex ensemble of multiple modes of representation orchestrated by the science teacher as a way of shaping scientific knowledge in the classroom. They argued that learning should not be seen as centrally dependent on oral and written language, but rather as a "dynamic process of transformative sign-making." In addition, they also pointed out the "semiotic affordances" of different modes of representation in realizing different kind of meanings. Lemke, in his analysis of canonical scientific texts (1998), also came to the same conclusion that scientific knowledge is constructed through joint meaning-making across multiple modes of representation. In particular, he highlighted that the possibility of making different meanings increases when multiple modes of representation are combined. This "multiplying meaning" effect is what makes possible the concepts of science to evolve historically as "semiotic hybrids of verbal, mathematical, visual-graphical, and actional-operational mode" (Lemke 1998, p. 87). Therefore, multimodal integration is an indispensable part of science learning as well as disciplinary literacy teaching in the sciences.

Education Context in Singapore

In Singapore, English is the medium of instruction for all subjects, except for the Mother Tongue languages. As such, the way that the English language is used by both teachers and students in the teaching and learning of content subjects plays an important role in the overall development of students' language and communication skills. In recent years, the Singapore Ministry of Education (MOE) launched an initiative to support the development of English competence and effective communication in all schools. Called the Whole School Approach to Effective Communication in English (WSA-EC), this program is a strategic endeavor to improve students' communication skills in English (ELIS 2011). The MOE has argued that a key aspect of developing students' effective communication in English should be the understanding that language competency applies to all subject classrooms, and should not be restricted to the English language arts (ELA) classrooms. However, it is recognized that as every academic subject has its own unique disciplinary literacy, effective communication by subject teachers involves the skilful use of disciplinary-specific language to help students better understand, process and internalize subject knowledge effectively.

The focus on subject-specific communication in WSA-EC is timely and relevant in reinforcing the focus areas of MOE's twenty-first century competencies framework that will prepare students for the demands of this century (MOE 2010). One of the core competencies in this national framework is communication skills: "Communicating effectively refers to the delivery of information and ideas coherently, in *multimodal* ways, for specific purposes, audiences, and contexts" (Standards and Benchmarks document for twenty-first century competencies, MOE 2011). Communication is conceptualized as the interactive process of sharing concepts, thoughts and feelings between people using the medium of language as a resource. In addition, this process involves the co-construction of meaning by those involved in communication. Effective communication occurs when the audience or reader understands a message in the way the communicator intended it to be understood, or when the co-construction of meaning satisfies all parties involved. Thus, communication is at the very heart of learning. Research into effective communication across the curriculum, given the emphasis on spoken, written, and multimodal communication skills and the need to explore interaction in subject classrooms is thus critical.

Research Context and Methodology

With the WSA-EC program providing a relevant context, the research project discussed in this chapter aimed to help two secondary schools in Singapore achieve the desired outcome of development of effective communication for science students. One of the objectives in this 3-year research project was to develop good disciplinary literacy teaching strategies with several collaborating teachers. To accomplish this, a design-based research approach (Collins et al. 2004) was adopted whereby pedagogical interventions, such as lesson activities and worksheets, were designed, implemented, evaluated, and refined through several iterative cycles. All instructional materials were co-developed by the researchers and the collaborating teachers based on a disciplinary literacy framework that we developed for this research project.

In this chapter, the lesson activities and worksheets presented were based on the first intervention cycle. The analysis also focused on two teachers – Derrick and Kathryn – from one of the collaborating schools. (All names are pseudonyms to protect privacy). Derrick was a physics teacher with 5 years of teaching experience while Kathryn was a chemistry teacher with 8 years of teaching experience. Both Derrick and Kathryn taught grade nine students from diverse ethnic groups comprising Chinese, Malay and Indian. The average class size was 28 students. The students were generally quiet but some could be active and vocal when presented with questions.

Ethnographic methods consisting of participant-observation, video recording, field-note taking, and artifact collection were used to collect data from the observed classrooms. The primary data source for the study reported in this chapter is classroom videos, comprising 12 lessons (11 h and 10 min in total) covering the topic of waves for physics and chemical bonding for chemistry. The videos were recorded by one camera at the back of the classroom focusing on the teacher. Another data source is students' writing on all the worksheets designed for the intervention research.

The analytical framework is based on a previous framework developed to examine the role of multiple and multimodal representations for science meaning-making (Tang et al. 2014). Multiple representations refer to the practice of re-representing the same phenomenon using different instructional resources, while multimodal representations refers to the use of multiple modes of representation (e.g., words, diagrams, graphs) to construct meaning. This framework incorporates the theoretical notion of "re-representation" (Hubber et al. 2010) and "transformative sign-making" (Kress et al. 2001) as the transformation of representations from one instructional resource to another across a series of activities in a lesson, as well as the notion of "semiotic affordances" from SFL (Kress et al. 2001) in examining the "multiplying meaning" effect (Lemke 1998) of combining multiple modes of representation. In addition, we also use the framework from Mortimer and Scott (2003) to analyze the various types of communicative mode (e.g., dialogic, authoritative, interactive, non-interactive) between the teacher and students. In particular, we focused on whether the teacher used a dialogic approach in considering multiple "voices" from the students (Bakhtin 1986) or an authoritative approach in considering only the scientific point of view.

Based on this analytical framework, lesson videos were viewed, coded, and analyzed using Transana software in two stages. The first stage involved the segmentation of the continuous sequences in a lesson video into meaningful discrete units. The boundaries of each segment are determined by the demarcation of prominent shifts occurring in the classroom, such as a discernible change in the participants' interaction pattern or the texts to which they are oriented. Each segment is then coded and tagged according to four categories: teaching activity (e.g., teacher explanation or group experiment), communicative mode (e.g., dialogic, authoritative; Mortimer and Scott 2003), instructional resource (e.g., video, worksheets; Hubber et al. 2010) and the mode of representation (e.g., words, diagrams; Kress et al. 2001; Lemke 1998). At this stage of analysis, the dialogue was not transcribed at this point due to the time-consuming nature of transcription. However, the analysis allowed us to construct the teaching sequence of every lesson and the corresponding communicative mode, use of resources, and mode of representation (see Appendix 1). This analysis also facilitated the identification of relevant episodes for the next stage of analysis.

The second stage of analysis involved an in-depth micro-analysis (Erickson 1992) of the teachers' and students' dialogue and multimodal interactions (e.g., gestures) and artifacts (e.g., students' writing). Spoken language was first transcribed and, together with written language, analyzed at the level of a clause

(Halliday 1978). The meaning of each clause was interpreted through the semantic relationship among the words in the clause. For instance, the clause "the water molecules are balanced" is an attributive relationship between a medium (water molecule) and its attribute. For visual image and gesture, a similar analysis was carried out using Kress and van Leeuwen's (1996) and Martinec's (2000) frameworks. Thus, the same example of "water molecules are balanced" is also realized visually through the drawing of circles (signifying the medium) and wavy lines (signifying the attribute) and gesturally through the left and right hands (signifying the medium) moving up and down alternatively (signifying the action of balancing). See these examples in the later analysis.

A Multimodal Approach to Disciplinary Literacy Teaching in Physics

The first example of a multimodal approach to disciplinary literacy teaching is based on two 1-hour physics lessons on the topic of waves. The overall lesson objective was for the students to describe the movement of particles and the transfer of energy in transverse wave motion. In particular, the students needed to discern that the particles in a wave vibrate about a fixed point instead of moving along with the forward propagation of the wave. From this distinction, the students were then asked to explain how energy can be transferred in a wave motion without the physical transfer of matter.

One month before the physics lessons took place, the researchers met with the teacher, Derrick, to discuss the lesson activities and design the worksheets to be used in the lesson. It was recognized from the discussion that many students tend to have difficulties observing the vibration of the wave particles amidst the dynamic fast-changing motion of a wave. Furthermore, to give a scientific description of the wave motion would entail a multimodal competency involving making connections between a series of disciplinary-specific diagrams and a set of scientific terminologies, such as vibration, kinetic energy, transfer, fixed position, medium, perpendicular, and adjacent particles. As such, a series of progressive activities was planned which involved the students doing a hands-on experiment with ropes at the beginning and writing an account of the wave motion by the end of the lesson. At the same time, attention was also given to the literacy activities in the lesson by planning several group discussions, and individual writing and multimodal integration exercises.

To illustrate the role of multimodal activities in the development of the students' scientific description of wave motion, three specific episodes will be analyzed and presented in this section. These episodes were selected primarily due to a notable shift in the communicative mode (from dialogic to authoritative and vice-versa) and/or the mode of representation (physical, visual, written text). In our analysis, it was found that rich multimodal integration often occurred during these shifts. Table 8.1 shows the teaching activity, communicative mode, instructional resource and

Table 8.1 The teaching activity, communicative mode, instructional resource and modes of representation for selected episodes

Video time	Teaching/learning activity	Communicative mode	Instructional resource	Mode of representation
Episode 1, Lesson 1				
1:00–8:59	Teacher introduces and demonstrates the rope experiment	Non-interactive/ authoritative	Rope with colored knots (see Fig. 8.1)	Physical
8:59–20:54	Students carry out rope experiment in pairs	Interactive/dialogic	Rope with colored knots	Physical
	Students discuss their observation in pairs and write individually on a given worksheet		Worksheet (page 1)	Written/visual (static diagrams)
Episode 2, Lesson 1				
41:24–46:27	Teacher discusses with students their responses	Interactive/dialogic		
47:43–52:40	Teacher summarizes the discussion and relates the video to the rope experiment	Interactive/ authoritative	Video Worksheet (page 1 and 2)	Visual (animated) Written/visual (static diagrams)
Episode 3, Lesson 2				
12:30–16:32	Teacher discusses with students their written responses from the last lesson	Interactive/dialogic	Worksheet (page 3)	Written/visual (static diagrams)
16:32–26:29	Teacher generalizes the structure and the sequences in the explanation	Non-interactive/ authoritative	Fill-in-the-blanks notes	Written
26:29–33:08	Students revise their explanations to the earlier questions		Worksheet (page 3)	Written/visual (static diagrams)

modes of representation for these three episodes. The entire teaching sequence for the two lessons is shown in Appendix 1.

Episode 1. From Physical Demonstration to Initial Observation

The lesson began with Derrick giving an overview of what the students would be doing in subsequent activities:

1 Derrick: We will do some activities first, using the rope that I placed at your bench. You work in pairs, then. While you are doing the activity, these are the key words to take note of. Okay, observe, to feel, and think how you describe. How you draw. These are the

things you consider when you do the activity. Then we move on to the discussion. When we do the activities and the discussion right, I will be gathering your feedback and views. But I won't be clarifying. I will help to consolidate first.

Derrick's initial instruction overtly pointed to the multimodal nature of the subsequent activities and then set the stage for further student activity. First, the students carried out a hands-on activity in pairs using a rope. The aim of this activity was to generate a physical representation of wave motion. From this physical representation, the students would then be asked to re-represent (Hubber et al. 2010) their sensory and kinesthetic experiences into words and diagrams in a worksheet; according to the teacher's instruction to "describe", use "key words" and "draw". The worksheet is designed to help the students make connections among the various modes of representations (see Figs. 8.2 and 8.3 for samples of students' completed worksheet). After the students re-represented the physical representation into words and pictures in their worksheets, they would then "move on to the discussion" where the teacher would gather their preliminary ideas about wave motion. In sum, the students would be going through successive activities of doing, writing, drawing, and talking. Six minutes later, Derrick demonstrated to the students how to generate the wave motion by vibrating one end of the rope resting on a table, while a student held firmly on the other end (see Fig. 8.1). After the brief demonstration, Derrick then gave further instructions on how to fill out the worksheet:

> 2   Derrick: [pointing at screen] Right, so you look at it. From the observation, there are some sequential diagrams below. Right, there are some diagrams here [pointing at screen]. What you want to observe is. ugh.. excuse me. You take note of position A and B. It can be any of the two colors. Then, use arrows to indicate direction of motion for A and B. Right, pay attention to any of the two colors. And the direction of their motion. Describe what you see and how you feel.

Derrick pointed to several parts of the worksheet (see Fig. 8.2) projected on the screen as he gave these instructions. First, he directed the students' attention to the sequence of diagrams in the left column of Fig. 8.2, where each diagram shows a snapshot of the wave motion at different times. He then asked the students to "take note of position A and B." In the rope experiment, four colored knots were tied onto the rope at intervals of approximately 10 cm apart. The purpose of these knots was to help the students observe that the movement of the knots is perpendicular to rather than along the rope. In the worksheet, the circle labelled A and B on the first diagram represented the positions of those colored knots. The students would then indicate, for subsequent diagrams, the positions of A and B as well as their direction of motion, according to their observation. In this way, the activity aided students in re-representing the positions and motion of the knots on the physical rope into circles and arrows on the worksheet. This initial re-representation from the demonstration (activity of doing) to the diagrams (activity of drawing) is an important first step in the multimodal integration process required in understanding wave motion.

After the completion of the diagram, the students were asked to "describe what [they] see and feel" on the worksheet. Figure 8.2 shows the completed worksheet from student Hidayah. Her writing at this stage revealed the emerging language she had for describing her observation of the wave motion. Hidayah only described the

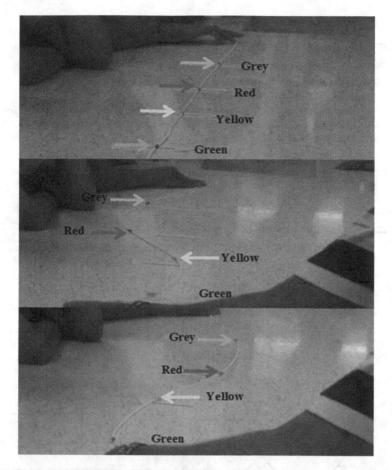

Fig. 8.1 Demonstration of wave motion using a rope by Derrick

wave motion in general (i.e., "the string seems to be doing a wave"). Her statement "The wave is transferred from A and B to the other points to the right" indicates her observation of the transfer of the wave motion in general, but not the specific motion of the particles, as represented by A and B. Furthermore, her sequential diagrams did not indicate the directions of motion of A and B. Thus, it was not clear whether this student could differentiate between the vibration of A and B (perpendicular to rope) and the direction of the wave motion (parallel to rope). This was a common challenge faced by many other students at this stage of the lesson.

At this stage, there are two major characteristics of Hidayah's worksheet that indicate she has not provided a complete description of the concept. First, she has not developed the language specificity to give an accurate account of wave motion in her writing. Second, there was little connection between her drawing and writing. In the next episode, we move forward in the lesson to look at how Derrick and the students, in talking about a water wave phenomenon, developed the necessary language and multimodal connection to give a scientific account of wave motion.

Part 1: Transverse waves

Experiment 1

1. In a pair, hold both ends of the rope on the table by pressing your fingers on them lightly.
2. Oscillate one end of the rope slowly and then increase the speed of oscillation.
3. Observe and feel what happens during the experiment.
4. From your observations, in the sequential diagrams below, diagrams, label the positions of A and B. Indicate using arrows the directions of motion of A and B and the wave.
5. Describe what you see and feel as the rope is oscillating.

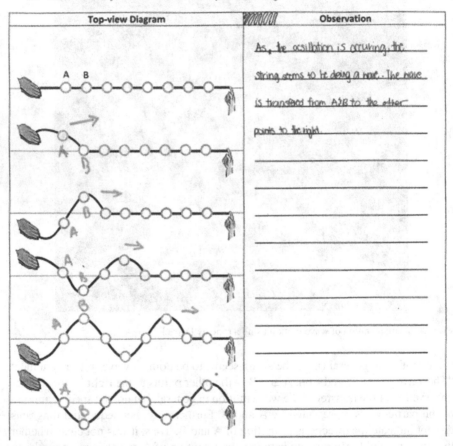

Fig. 8.2 Page 1 of Hidayah's completed worksheet

Episode 2. Refining Language from Video Observation

After most students have completed page 1 of the worksheet (i.e., Fig. 8.2), Derrick went on to discuss some of the students' writing. He picked a few students' work and showed them to the class using a document camera. As stated by Derrick at the beginning of the lesson, his purpose at this point was only to consolidate the students' initial responses. He refrained from providing the "model answer" to the students.

Instead of giving an authoritative answer to the question of how to describe wave motion, Derrick used a more dialogic approach (Mortimer and Scott 2003) to elicit and discuss students' views concerning the motion of the particles in relation to the wave propagation. Halfway through the lesson, Derrick showed a video of a water wave passing from left to right and a ball floating up and down on the surface of the water wave. The students were then given time to discuss the question of "what is moving in a wave motion, and in what direction?" The following transcript shows one of the pivotal moments in the discussion between Derrick and the class:

[Lesson 1, Time: 42:27]

1 Derrick: … Audrey? What did you all discuss?
2 Audrey: The particles don't move sideways, they move up and down.
3 Derrick: Particles don't move sideways, they move up and down. When you talk about particles, what particles are you talking about?
4 Audrey: (inaudible)
5 Derrick: When you refer to particles, what particles are you talking about?
6 Audrey: The particles in the water
7 Derrick: The particles in the water, are you talking about water molecules? I mean, particle is nothing wrong. I just want to be more specific. Yah? (pointing at Amu)
8 Amu: The ball is not moving because the particles in the liquid is.. (inaudible)
9 Derrick: The ball is not moving because the particle in the liquid is?
10 Amu: Is balanced. Because it's moving. The particles are, the water molecules are moving.. (left and right hands gesturing a vibration motion)
11 Derrick: Balanced. Alright. Okay, so that is some idea you have in mind.. Alright. Who else?.. Rui En? What did you all discuss?
12 Rui En: It's like erm.,
13 Derrick: Quite interesting to hear.
14 Rui En: It's like the.. what's that call ah? The wave molecules, right? The water molecules.
15 Derrick: Water molecules, uh huh.
16 Rui En: Balanced.
17 Derrick: Balanced. Meaning?
18 Hwee Ling: It's like when one is like..
19 Rui En: It's up, then the other is down
20 Derrick: So water molecule is balanced when one is up, the other one is down.
21 Rui En: Yah, so it is like pushing. So like it stays put..
22 Derrick: So it's pushing what?
23 Rui En: Cause like one is up and one is down, right? (Left hand raises up while right hand moves down). Then the base is here right? (Points at the middle). Then it like it goes here So like..
24 Derrick: More like a see-saw, is it?
25 Rui En: Yah
26 Derrick: Like a see-saw, you, you..
27 Rui En: Something like that
28 Derrick: Is that what you are saying? Okay. So she's saying that it is more like a see-saw. When one is moving up (left hand raises up), one is moving down (left hand moves down), so it would just be just rocking left and right (right hand animates a waving action.

There are two important insights regarding multimodal integration from this episode. The first insight is the joint refinement of ideas that developed in tandem with the increasing specificity of the language used by the speakers. In line 2, when Audrey was talking about the movement of the particles, Derrick asked her to specify what particles she was referring to (line 3, 5). Through Audrey's response in line 6 and Derrick's paraphrasing in line 7, they began to use the more specific term "water molecules" instead of the generic "particles." This term was subsequently appropriated by other students. For instance, Amu consciously changed "particles" to "water molecules" in line 10, and Rui En took a while to recall the correct term to use in line 14. This refinement of the language is important as it allowed the teacher and students to distinguish between the water molecules (as carrier) and the ball (as object being carried) in the subsequent discussion. Only through this distinction can we understand the nuanced meaning of the word "balanced", which was first brought up by Amu (line 10) and later elaborated by Rui En and Hwee Ling (line 16–27).

To summarize the discussion, what Rui En meant by the "water molecules are balanced" (line 14–16, 20–21) is that the two sides of the water surface surrounding the ball are alternatively moving up and down (as seen from the gestural action of her left and right hands in line 23). Because of this "balanced" motion – analogous to a see-saw motion as suggested by Derrick (line 24) – the ball will not be carried along with the water wave. This was indirectly stated in line 21 when Rui En said "it [the ball] stays put." Derrick, in line 28, tried to summarize Rui En's idea by giving the analogy of a see-saw. He also reiterated that the water molecules are moving up and down, while the ball is rocking left and right due to the see-saw motion of the surrounding water molecules. It must be noted that there is some ambiguity in Derrick's summary in line 28 due to the imprecise use of the pronouns ("one", "it") to indicate different aspects of the observation: the water wave motion, the water molecules, and the ball. Nevertheless, what is crucial in this discussion is the consensus that the water molecules are moving up and down, rather than left to right, along the direction of the water wave. Furthermore, the teacher and students recognized that this distinction accounts for why the ball does not travel along with the water wave.

The second insight we can gain from this episode is the actions taken by Derrick to explicitly support the students in developing the language and multimodal integration competency in order to accurately describe the wave motion from a physics viewpoint. It has already been demonstrated how Derrick aided the students in the use of "more specific" terms to describe the wave in line 7 and to clarify what students meant by the word "balanced" in line 17. Another important aspect of the teacher's action was to help students make the connection between what they saw in the video and how they described the animated motion. The following excerpt provides an instance where this multimodal connection was made directly by Derrick. This excerpt occurred about 3 min after the last discussion. Derrick paused at a particular scene on the video and asked the following question:

[Lesson 1, Time: 47:43]

29 Derrick: (Pointing at the screen) Which direction is the wave traveling?
30 Class: Right to left.
31 Derrick: Right to left. Alright, right to left ah. (Pointing at a particular point on screen) Which direction is this point traveling?
32 Class: Up and down.
33 Derrick: Is it moving along with it? (Gesturing from right to left)
34 (Several students shook their heads)
35 Derrick: No. Turn back to the first page. Did you indicate that? That the particle A B is moving up and down? Did you indicate that the particle is moving up and down or have you actually indicated that it has moving toward the right? Does that clarify how the particle is moving?

At this point, Derrick was doing a summary to round up the earlier discussion. Unlike the earlier part of the discussion where he was using a dialogic approach to elicit various viewpoints, there is a notable shift toward a more authoritative approach (Mortimer and Scott 2003) where he channeled the discussion toward the scientific description. This could be seen from the highly distinguishable I-R-E (Initiate-Response-Evaluate) pattern in this discussion (Mehan 1979). At the same time, he used this authoritative approach in conjunction with the visual scenes in the video to guide the students in discerning the right-to-left and up-down motion of the traveling wave and vibrating particle respectively. In line 35, Derrick then made a very important move, which was to link the discussion back to the rope experiment the students had earlier completed on "the first page" (line 35) of the worksheet, as shown in Fig. 8.2. In particular, Derrick directed the students to differentiate the up-down vibration of points A and B in contrast to the left-right movement of the wave motion. This multimodal connection across the specific activities of doing (the rope experiment), drawing/writing (Fig. 8.2), talking (episode 2), and viewing (video in episode 2) is an important aspect of the disciplinary literacy teaching observed in this lesson.

Episode 3. Explaining Energy Transfer Using More Refined Language

Near the end of the first lesson, Derrick signaled a change in the focus toward talking about energy transfer in a wave motion. In particular, the students were instructed to fill out page 3 of the worksheet. This part of the worksheet was designed to help them "explain how energy is transferred in a ripple produced by dropping a pebble into a pond in terms of the motion of the water particle" (see Fig. 8.3).

Earlier, the quality of Hidayah's description of the wave motion was examined (see Fig. 8.2) and it was indicated that she lacked the language specificity to give an accurate account of wave motion. This part of the worksheet shows her writing after the intervening discussion we described in episode 2. In Fig. 8.3, the words in bold and black were written by Hidayah at the end of the first lesson. Comparing her writing in Fig. 8.2 with that in Fig. 8.3, there are notable changes in how she was

Practice question

Explain how energy is transferred in a ripple produced by dropping a pebble into a pond in terms of the motion of the water particle. Use the diagrams provided below to explain your answer.

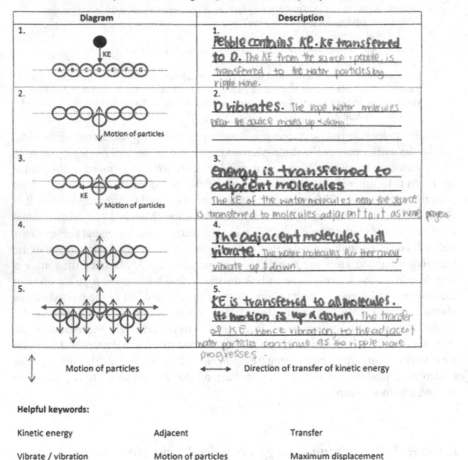

Diagram	Description
1.	1. Pebble contains KP. KE transferred to D. The KE from the source : pebble, is transferred to the water particles by ripple wave.
2.	2. D vibrates. The ripple water molecules near the source moves up & down.
3.	3. energy is transferred to adjacent molecules The KE of the water molecules near the source is transferred to molecules adjacent to it as wave progress
4.	4. The adjacent molecules will vibrate. The water molecules for the away vibrate up & down.
5.	5. KE is transferred to all molecules. Its motion is up & down. The transfer of KE, hence vibration, to the adjacent water particles continue as the ripple wave progresses.

↕ Motion of particles   ↔ Direction of transfer of kinetic energy

Helpful keywords:

Kinetic energy	Adjacent	Transfer
Vibrate / vibration	Motion of particles	Maximum displacement

Fig. 8.3 Page 3 of Hidayah's completed worksheet

able to distinguish the vertical motion of the water molecules in relation to the horizontal motion of the ripple wave. There is also an increased specificity in the language in terms of identifying different parts of the wave (e.g., adjacent molecules, all molecules) as well as the direction of motion (e.g., up, down). Part of this change could be due to Derrick's facilitated discussion described in episode 2. Another factor is likely the design of the worksheet in terms of the sequential diagrams placed next to the writing and the "useful keywords" provided such as "vibrate", "adjacent", and "transfer". Although these keywords were provided, it is important to note that the students' understanding of these words also depends on the discussion that was facilitated by Derrick. Thus, each of these methods is an important disciplinary literacy strategy that has helped Hidayah and other students produce a more scientific account of wave motion.

In the second lesson, Derrick discussed the students' writing on page 3 of the worksheet. He chose a few students' work and showed them to the class. Hidayah's worksheet was one of those he chose to discuss. In the following excerpt, Derrick read Hidayah's writing in line 1 and went on to refine her language in line 2:

[Lesson 2. Time: 12:30]

1   Derrick: Pebble contains KE. KE transferred to particle D. Particle D vibrates. Energy is transferred to adjacent molecules. Adjacent molecules vibrate. KE transferred to all molecules. Motion is up and down. Alright. Alright with the sequence?

2   Derrick: Um, possible choice of words, okay you can reconsider. It says pebble contains KE. Now I think we don't usually use the word contains right? Probably the word possess. Pebble possesses KE. D vibrates. But vibrates we want to be a bit more specific. In what direction? In what way? Because vibrate can be what? Circular? Swinging? Left, right? Up, down? Alright. Can be a bit more specific.

After Derrick discussed samples of students' writing and went through their explanations, he then proceeded to break down the required explanation into several parts in order to help the students analyze the structure and sequential steps in this explanation. About 10 min later, he gave the students time to revise and rewrite their explanation on page 3 of the worksheet. In Fig. 8.3, the words in blue were the revisions written by Hidayah after this part of the lesson. Again, comparing her revision with her earlier writing (in bold and black), we can see further improvement in her language. For instance, she has replaced the phrase "contains KE" with a more disciplinary appropriate phrase – "KE from the pebble".

A Multimodal Approach to Disciplinary Literacy Teaching in Chemistry

The second example of a multimodal approach to disciplinary literacy teaching in chemistry is presented here to provide a contrasting case to the earlier example in physics. However, due to space constraints, we will only narrate the key episodes without the supporting excerpts and analysis.

In these six 1-hour lessons, the overall lesson objective was to understand chemical bonding and the properties of each type of chemical substance. At the end of the lesson series, students were expected to be able to explain the properties each chemical substance exhibits. Based on a discussion with Kathryn prior to the lesson series, it was determined that students had difficulty discerning the chemical bonds in simple covalent substances and understanding how some substances only exhibit electrical conductivity in particular conditions. With this in mind, a lesson series was designed that required students not only to write but also to draw diagrams. The multimodal integration of written and visual representations was a necessity in the design of the lesson worksheet as we predicted that drawing diagrams would help students visualize and understand bonding between particles, and eventually write appropriate explanations of the properties exhibited by the substances. Literacy activities such as student discussion were also planned in the design of the lesson.

Episode 1: Drawing to Learn

In a lesson prior to this episode, the chemistry teacher, Kathryn, showed a video of an experiment designed to test the conductivity of an ionic compound, sodium chloride (table salt). In the video, when a circuit was connected across solid sodium chloride, a light bulb did not light up, thus demonstrating the non-conductivity of solid sodium chloride. However, when the solid sodium chloride in the set-up was heated and melted, the bulb lit up. Kathryn used this video as a context to teach chemical bonding and the properties of ionic and covalent compound. This video showed a macroscopic representation (Treagust et al. 2003) of the phenomenon, which is the illumination of the light bulb as seen by our naked eyes. To give a scientific account of why different solutions can conduct electricity, students would need to give a microscopic representation of the chemical substance in terms of the ions and electrons, which are invisible to us. Subsequently, Kathryn asked students to imagine the same experiment but with the solid sodium chloride replaced with ice made of de-ionized water, and to predict the outcome of the new experiment. Kathryn nominated a student Zhiwen to share her thoughts. Zhiwen predicted that neither the ice nor the melted ice could conduct electricity and, hence, light the light bulb due to the absence of charged particles. However, she could not elaborate why there was no charged particles.

Kathryn saw Zhiwen's difficulty in explaining the absence of charged particles and took the opportunity to get students to draw a microscopic diagram to show that there was indeed no charged particle in the ice and melted ice. This re-representation from a verbal to a visual mode taken by Kathryn was necessary and crucial to students' understanding of the nature of simple covalent substances. Unlike ionic substances whose bonds are broken upon heating, covalent bonds in simple covalent substances are not broken upon heating but their intermolecular forces of attraction are. This is the most common misconception in this sub-topic that Kathryn was hoping to address by having her students make drawings.

Despite having established that water was H_2O and despite the fact a diagram of ice (solid H_2O) was provided in the worksheet (Fig. 8.4), there were still students who drew the microscopic representation of water inaccurately. Figure 8.4 shows a diagram drawn by a student, Melissa. Following the diagram of ice provided, she represented the oxygen atoms as white circles while she represented the hydrogen atoms as black circles. However, instead of keeping the water molecules intact upon heating by drawing two black circles attached to one white circle, she drew the circles separately, suggesting the breaking up of strong covalent bonds. From her diagram, we could infer that Melissa might have difficulty in recognizing covalent bonds and intermolecular forces of attraction, and in understanding which forces of

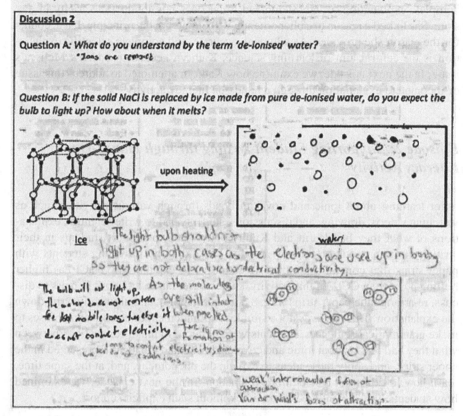

Fig. 8.4 Page 3 of Melissa's worksheet

attraction are broken upon heating simple covalent substances. Kathryn highlighted to students that such a drawing implied that "if (we) heat it (water) until gas, maybe it will split into electrons, and neutrons, and protons" which is inaccurate. Kathryn then drew the more accurate representation on the whiteboard and Melissa redrew her representation below the initial one as shown in Fig. 8.4.

After establishing that only intermolecular forces are broken upon heating ice, Kathryn instructed students to write an explanation why water cannot conduct electricity in any state with reference to the light bulb experiment they had watched. In Melissa's writing, she explained that "The bulb will not light up, the water does not contain free mobile ions; therefore it does not conduct electricity." In this explanation, she merely stated the condition of the water that it had an absence of mobile charged particle (e.g., ions) but failed to account for the absence by indicating that electrons are all used up for bonding and ions are already removed. This is an incomplete explanation as compared to the suggested answer given by Kathryn (shown in Fig. 8.4 in lighter ink).

Melissa's writing suggested that she may have had difficulty in constructing a complete explanation of why deionized water does not conduct electricity in any state. She linked a reason to a phenomenon without giving the principle that accounts for the reason she provided. As a logical, accurate, and complete scientific explanation is the central aim of learning science, Kathryn needed to address Melissa's issue. In the next episode, we examine how Kathryn attempted to address this issue of incomplete explanation through a hands-on literacy activity.

Episode 2: Supporting Student Writing through Hands-On Literacy Activity

After learning about ionic and covalent bonds through various activities such as watching videos, drawing, and discussions, students had to write scientific explanations of what they had learnt and Kathryn attempted to support students in their writing through hands-on literacy activity. She did so by providing students with paper strips that contain a jumbled-up sequence to explain why NaCl has higher boiling point than CCl$_4$. In this activity, students were asked to work in pairs to discuss, re-arrange the paper strips (Fig. 8.5 as an example), and ultimately write down the explanation. This task was not a simple re-arrangement of clauses and phrases to make grammatical sentences. Students were required to make connections between what they had learnt about ionic and covalent bonds and the ideas represented in the paper strips, and apply those ideas to explain the phenomena, and, at the same time, learn how to construct a scientific explanation. In the next episode, we examined how students write scientific explanation without such explicit support.

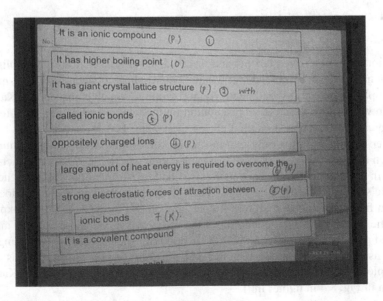

Fig. 8.5 Student's arrangement of the jumbled-up explanation sequence

Episode 3: Writing Without Explicit Support

Unlike the earlier episode, Kathryn did not give students explicit support in the form of strips of paper. Instead, she only reminded students of the writing structure that they had learnt earlier through using the strips of paper. This episode highlights the removal of the multimodal scaffold that Kathryn had previously provided students. Students were expected to be able to construct their own mental scaffold before writing down their answer.

In examining Melissa's answer (Fig. 8.6), it appears Melissa improved in terms of writing a thorough and complete, as well as scientifically accurate, explanation. Even without explicit scaffolding, Melissa was able to write a complete explanation. She did not merely state the reason that the heat energy required to break the weak intermolecular forces is low, but also gave the principle behind this reason, that is, simple covalent substances such as iodine have strong covalent bonds within the molecules but weak intermolecular forces between them. She also used the logical connective "as a result" to show a cause-and-effect relationship between the fact that the weak intermolecular forces can easily be overcome and thus, the low boiling point results.

Based on this writing, the multimodal scaffoldings in the forms of drawing and the strips of paper appear to have had a positive impact on Melissa's understanding

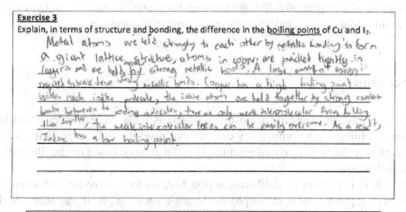

Fig. 8.6 Melissa's answer in exercise 3

of covalent compounds as well as her ability to construct a good scientific explanation. She has demonstrated understanding that strong covalent bonds are between the atoms within a simple covalent substance while weak intermolecular forces only hold the molecules together. Her writing also reflects that she is able to not only state the reason (heat energy requirement) but also the principle behind boiling point (bonding and forces).

The actions taken by Kathryn in scaffolding students' scientific explanation in this series of episodes highlight some potential uses of multimodal integration in disciplinary literacy teaching. First, Kathryn used the video of melting NaCl projected in the first lesson as the context of the question. This helped students identify which points in the strips of paper belong to NaCl, for example, the strips that describe high boiling points. Second, Kathryn gave students strips of paper as a visual aid in constructing the explanation as each of the strips represents an idea or a point. This encourages students to think of all the possible points discretely and then synthesize them in one explanation, just like rearranging the strips of paper. The multimodal integration across various activities utilized by Kathryn in teaching in this episode highlights an important aspect central to disciplinary literacy teaching.

However, we note that future improvement in this lesson series is to get students to provide diagrams in this explanation. The purpose is for them to further illustrate (visually) what they meant by "a giant lattice structure", "atoms in copper are packed tightly", "iodine atoms are held together by strong covalent bonds." This requirement of a multimodal linkage between texts and diagrams would strengthen their multimodal competency in writing a scientific explanation.

Conclusion

The main question emerging from this analysis is what we can infer about how multimodal integration competency can be developed through an explicit teaching of disciplinary literacy. Two major findings can be gleaned based on the illustrations given in this chapter.

First, it is important to deliberately plan lessons to involve several sequential stages of re-representation. In the physics lessons, for example, it was observed that Derrick planned a series of multimodal activities from a hands-on rope experiment to a group discussion to a visually-scaffolded writing exercise. As illustrated in the three episodes, the re-representation from an activity of doing to talking, drawing, and writing guided the students toward accurately explaining the wave motion. A similar sequence of re-representations was observed in Kathryn's case, although her lesson did not have a "doing" activity at the beginning. Instead, this was replaced by a video which showed macroscopically the outcome of an experiment – the lighting of a light bulb based on a liquid's conductivity. Through sequential activities of talking, drawing, and writing, the students were guided to explain in a microscopic sense the reasoning behind the experiment.

Second, during lesson implementation, teachers need to be very explicit in pointing out to the students the specific language and multimodal connections that are required in the scientific explanations. This attention to the language and representation specificity can occur during the shift from a dialogic communicative mode where the students' ideas were elicited to an authoritative mode where the discussion became more directed by the teacher. During this shift, we saw from both Derrick's and Kathryn's lessons how they guided the students to (i) use more specific terms (e.g., water molecules, de-ionized), (ii) link what they saw in the videos to their verbal explanation, (iii) make corrections of their diagrams, and (iv) visualize and discuss the logical sequence of their written explanation. Such actions taken by the teachers are necessary in developing the students' competency in using multimodal representations in their scientific explanations.

These two findings provide insights for researchers and teachers on how to design and carry out multimodal activities as part of the focus on disciplinary literacy teaching to develop students' competencies in multimodal communication. Although this research project started with an awareness of the importance of multiple and multimodal representations, we did not have a clear idea of how to translate the theoretical ideas into classroom practices within the context of the Singapore educational system. Thus, the rich description of the teaching sequences presented in this chapter was aimed to provide exemplars for educators to learn as well as to replicate or modify. In this respect, the findings and analysis here provide a starting point that will inform us of the next phase of our research. In particular, we will continue to extend our findings to other lesson observations and develop a pedagogical framework or guiding principles that could inform teachers on the design and implementation of multimodal activities to teach disciplinary literacy in science.

In sum, past research has separately shown the importance of giving students the opportunity to talk and write in science classrooms. While each literacy activity is crucial to science learning, we argue, in this study, that it is equally important for each of these talking and writing activities to be connected not just to each other, but also holistically to other activities of doing and drawing. The integration of these multimodal literacy activities of talking, writing, drawing and doing is an important aspect of what we see as *disciplinary literacy teaching* in science.

Acknowledgement This chapter refers to data from the research project "Developing Disciplinary Literacy Pedagogy in the Sciences" (OER 48/12 TKS), funded by the Education Research Funding Programme, National Institute of Education (NIE), Nanyang Technological University, Singapore. The views expressed in this paper are the authors' and do not necessarily represent the views of NIE.

Appendix 1

Teaching sequence and corresponding communicative mode, use of resources, and modes of representation for the physics lessons. Selected segments shown in the analysis are shaded in grey.

Video Time	Teaching/Learning Activity	Dominant Communicative Mode (Mortimer & Scott, 2003)	Instructional Resource (Hubber et al, 2010)	Mode of Representation (Kress et al, 2001; Lemke, 1998)
Lesson 1				
1:00 – 8:59	Teacher introduces and demonstrates the rope experiment	Non-interactive/ Authoritative	Rope with colored knots (see Figure 8.1)	Physical
8:59 – 20:54	Students carry out rope experiment in pairs	Interactive/ Dialogic	Rope with colored knots	Physical
	Students discuss their observation in pairs and write individually on a given worksheet	Interactive/ Dialogic	Worksheet (page 1)	Written/Visual (static diagrams)
20:54 – 36:53	Teacher discusses with students their preliminary observations	Interactive/ Dialogic	Worksheet (page 1 and 2)	Written/Visual (static diagrams)
36:52 – 38:59	Teacher plays a video of water wave motion	Non-interactive/ Authoritative	Video	Visual (animated)
38:59 – 41:24	Students discuss in pairs the question: "In a wave motion, what is moving and in what direction?"	Interactive/ Dialogic		
41:24 – 46:27	Teacher discusses with students their responses	Interactive/ Dialogic		
47:43 – 52:40	Teacher summarizes the discussion and relates the video to the rope experiment	Interactive/ Authoritative	Video	Visual (animated)
			Worksheet (page 1 and 2)	Written/Visual (static diagrams)
52:40 – 57:14	Students write an explanation of "how energy is transferred in a ripple in terms of the motion of the particles"	Non-interactive/ Dialogic	Worksheet (page 3)	Written/Visual (static diagrams)
Lesson 2				
2:38 – 12:30	Teacher recaps last lesson and introduces key terms of wave motion	Non-interactive/ Authoritative		
12:30 – 16:32	Teacher discusses with students their written responses from the last lesson	Interactive/ dialogic	Worksheet (page 3)	Written/Visual (static diagrams)
16:32 – 26:29	Teacher generalizes the structure and the sequences in the explanation	Non-interactive/ Authoritative	Fill-in-the-blanks notes	Written
26:29 – 33:08	Students revise their explanations to the earlier questions	Non-interactive/ Authoritative	Worksheet (page 3)	Written/Visual (static diagrams)
33:08 – 42:56	Teacher discusses the solutions with the class	Interactive/ Authoritative		
42:56 – 53:09	Students attempt last question on their worksheet	Non-interactive/ Authoritative	Worksheet (page 4)	Written/Visual (static diagrams)

References

Bakhtin, M. M. (1986). *Speech genres and other late essays* (1st ed.). Austin: University of Texas Press.

Collins, A., Joseph, D., & Bielaczyc, K. (2004). Design research: Theoretical & methodological Issues. *The Journal of the Learning Sciences, 13*(1), 15–42.

Council of Chief State School Officers. (2010). *Common core state standards*. Washington, DC: National Governors Association Center for Best Practices, Council of Chief State School Officers.

English Language Institute of Singapore (ELIS). (2011). Whole school approach to effective communication. http://www.elis.moe.edu.sg/professional-learning/subject-literacy

Erickson, F. (1992). Ethnographic microanalysis of interaction. In M. D. LeCompte, W. L. Millroy, & J. Preissle (Eds.), The handbook of qualitative research in education (pp. 201–225). San Diego: Academic Press.

Halliday, M. A. K. (1978). *Language as social semiotic: The social interpretation of language and meaning*. London: Edward Arnold.

Halliday, M. A. K., & Martin, J. R. (1993). *Writing science: Literacy and discursive power.* Pittsburgh: University of Pittsburgh Press.

Hand, B., Lawrence, C., & Yore, L. D. (1999). A writing in science framework designed to enhance science literacy. *International Journal of Science Education, 21*(10), 1021–1035.

Hubber, P., Tytler, R., & Haslam, F. (2010). Teaching and learning about force with a representational focus: Pedagogy and teacher change. *Research in Science Education, 40*(1), 5–28.

Jewitt, C. (2008). Multimodality and literacy in school classrooms. Review of Research in Education, 32, 241–267.

Kelly, G. J. (2008). Learning science: Discursive practices. In N. H. Hornberger (Ed.), *Encyclopedia of language and education*. Boston: Springer Science + Business Media LLC.

Kress, G., & van Leeuwen, T. (1996). *Reading images: The grammar of visual design*. London/New York: Routledge.

Kress, G., Jewitt, C., Ogborn, J., & Tsatsarelis, C. (2001). *Multimodal teaching and learning: The rhetorics of the science classroom*. London: Continuum.

Lemke, J. L. (1990). *Talking science: Language, learning and values*. Norwood: Ablex.

Lemke, J. L. (1998). Multiplying meaning: Visual and verbal semiotics in scientific text. In J. Martin & R. Veel (Eds.), *Reading science*. London/New York: Routledge.

Martinec, R. (2000). Types of process in action. *Semiotica, 130*(3–4), 243–268.

McConachie, S., Hall, M., Resnick, L., Raci, A., Bill, V., Bintz, J., et al. (2006). Task, text, and talk: Literacy for all subjects. *Educational Leadership, 64*(1), 8–14.

Mehan, H. (1979). *Learning lessons: Social organization in the classroom*. Cambridge, MA: Harvard University Press.

Ministry of Education. (2010). MOE to enhance learning of 21st century competencies and strengthen art, music and physical education. Press release, 9 March. http://www.moe.gov.sg/media/press/2010/03/moe-to-enhance-learning-of-21s.php

Ministry of Education. (2011). Standards and benchmarks for 21st century competencies. Singapore: Curriculum Policy Office, Ministry of Education.

Moje, E. B. (2007). Developing socially just subject-matter instruction: A review of the literature on disciplinary literacy teaching. *Review of Research in Education, 31*, 1–44.

Mortimer, E. F., & Scott, P. (2003). *Meaning making in secondary science classrooms*. Buckingham: Open University Press.

Norris, S. P., & Phillips, L. M. (2003). How literacy in its fundamental sense is central to scientific literacy. *Science Education, 87*(2), 224–240.

Schleppegrell, M. (2004). *The language of schooling: A functional linguistics perspective*. Mahwah: Lawrence Erlbaum Associates.

Tang, K. S. (2011). Reassembling curricular concepts: A multimodal approach to the study of curriculum and instruction. *International Journal of Science and Mathematics Education, 9*, 109–135.

Tang, K. S., Delgado, C., & Moje, E. B. (2014). An integrative framework for the analysis of multiple and multimodal representations for meaning-making in science education. *Science Education, 98*(2), 305–326.

Treagust, D., Chittleborough, G., & Mamiala, T. (2003). The role of submicroscopic and symbolic representations in chemical explanations. *International Journal of Science Education, 25*(11), 1353–1368.

Tytler, R., Peterson, S., & Prain, V. (2006). Picturing evaporation: Learning science literacy through a particle representation. *Teaching Science, the Journal of the Australian Science Teachers Association, 52*(1), 12–17.

Unsworth, L. (2001). Evaluating the language of different types of explanations in junior high school science texts. *International Journal of Science Education, 23*(6), 585–609.

Veel, R. (1997). Learning how to mean-scientifically speaking: Apprenticeship into scientific discourse in the secondary school. In C. Frances & J. Martin (Eds.), *Genre and institutions: Social processes in the workplace and school* (pp. 161–195). London: Cassell.

Yore, L. D., & Shymansky, J. A. (1985). *Reading, understanding, remembering and using information in written science materials.* ERIC Document Reproduction Service No. ED 258825.

Chapter 9
Constructing Representations to Learn Science

Russell Tytler and Peter Hubber

Introduction: The Theoretical Basis of Our Work

There is a substantial literature, set broadly within a socio cultural framework, arguing that learning and knowing should be seen as a process of enculturation into the discursive practices of science (Lave and Wenger 1991), and further that these practices are substantially shaped around a set of discipline specific and generic literacies used in science to build and validate knowledge (Moje 2007). From this perspective, understanding is seen in terms of the capacity to generate and coordinate multi-modal representational resources to develop explanations and solve problems. Thus, explicit discussion of the form and function of scientific representations becomes a key aspect of teaching and learning in science (Ainsworth 2006, 2008; Lemke 2004). Achieving meta-representational competence (DiSessa 2004) in recognizing the characteristics of effective representational practice, including representational quality, the selective nature of representations, and how they are coordinated in developing solutions, becomes a key aim of science education (Gilbert 2005; Kozma and Russell 1997, 2005; Kozma et al. 2000). A growing modeling literature identifies the power of refinement of explanatory models through classroom negotiation (Clement and Rea-Ramirez 2008). The implications of these literatures for classroom teaching and learning practice is that there should be represented in the public space of the classroom the opportunity for students to develop a) explicit knowledge of representational form and function, b) knowledge of representational quality and the selective nature of representations, and c) skills in coordinating multiple representations in problem solving.

Further studies have verified the defining, rather than supporting role played by representations in the generation of knowledge and the solving of problems (Klein

R. Tytler (✉) • P. Hubber
Deakin University, Burwood, Australia
e-mail: russell.tytler@deakin.edu.au

© Springer International Publishing Switzerland 2016   159
B. Hand et al. (eds.), *Using Multimodal Representations to Support Learning in the Science Classroom*, DOI 10.1007/978-3-319-16450-2_9

2000; Tytler et al. 2009; Zhang 1997), consistent with pragmatist perspectives on the material nature of knowledge (Peirce 1931/1958; Wittgenstein 1972) where a concept is understood only by experiencing the effects of applying the representational practices constituting it to meaningful practical settings. Again, from a classroom perspective this implies that rather than focus on concepts conceived of traditionally as explicitly and verbally defined, teaching and learning processes need to be built around the representational resources used to instantiate scientific concepts and practices (Moje 2007).

This focus on representation work in science mirrors insights into the material basis of knowledge production in science itself (Pickering 1995; Gooding 2006). Digital, material and symbolic representations are crucial tools for reasoning in science, rather than simply records of abstracted and resolved ideas and theories (Gooding 2006). In generating and using representations, scientists reason not only through formal logic but also informal processes such as visual and spatial pattern recognition and metaphoric association (Gooding 2006; Klein 2006). Gooding's (2004, 2006) analyses of Michael Faraday's diaries demonstrate the complex process of representational re-description through which Faraday selected, abstracted and integrated effects used diagrams, physical apparatus, and complex reasoning processes to select and abstract, in establishing new knowledge. Studies of contemporary science teams and laboratory practices (Latour 1999; Nersessian 2008) show how models are constructed through elaborate cycles of data generation and re-representation (Nersessian 2008), to develop explanations and theories.

From socio-cultural perspectives, learners need to participate in authentic activities with these cultural resources/tools to learn effectively (Cole and Wertsch 1996; Vygotsky 1981). Other literature in science education argues the need for students to actively construct representations in order to become competent in scientific practices and to learn through participating in the reasoning processes of science (Ford and Forman 2006). Socio-cultural accounts of the value of this practice focus on the potential for increased student engagement in a learning community (Greeno 2009; Kozma and Russell 2005). From a cognitive perspective, Bransford and Schwartz (1999) sought to re-conceptualize the learning gains and potential for transfer when students generated their own representations. Rather than argue that students developed transferable domain knowledge from this activity, they claimed that student construction of representations led to the development of problem-solving skills that could be applied in new contexts. Cazden (1981) argued that students needed to engage with performance of representational practice before achieving competence.

Researchers in classroom studies in this area (Cox 1999; Greeno and Hall 1997; Lehrer and Schauble 2006; Tytler et al. 2006; Waldrip et al. 2010) have noted the importance of teacher and student negotiation of the meanings evident in verbal, visual, mathematical and gestural representations in science. They claimed that students benefited from multiple opportunities to explore, engage, elaborate and re-represent ongoing understandings in the same and different representations. Greeno and Hall (1997) argued that different forms of representation supported contrasting understanding of topics, and that students needed to explore the advantages and

limitations of particular representations. These studies also indicated that representations in science classrooms can serve many different purposes. As noted by Cox
(1999) the representations can be used as tools for many different forms of reasoning such as for initial, speculative thinking, to record observations, to show a
sequence or process in time (Ainsworth et al. 2011), to sort information, or predict
outcomes. Students need to learn how to select appropriate representations for
addressing particular needs, and be able to judge their effectiveness in achieving
particular purposes.

A major strand in theorizing learning processes in science involves research on
student model-based reasoning through inquiry (Clement 2000; Gilbert 2005;
Lehrer and Schauble 2006; Vosniadou 1994). Advocates of this approach argue that
the process of constructing, critiquing, testing and revising models is the key mechanism for promoting student conceptual growth. A growing modeling literature
identifies the power of refinement of explanatory models through classroom negotiation to achieve quality learning (Clement and Rea-Ramirez 2008; Manz 2012).
We have articulated the processes by which the construction and use of representations serve to frame both knowledge building and learning in science, through the
construct of affordances (Gibson 1979; Prain and Tytler 2012, 2013). We argue that
for students constructing a representation in a particular mode their attention will be
channeled in productive ways by the constraints inherent in that mode. Thus, visual,
textual, embodied (gesture and object manipulation) and symbolic modes have specific affordances that constrain what is observed and interpreted, and conceptual
understanding is built through the coordination of these multiple representations
each of which is partial, and approximate.

Thus, the literature informing our practice has emphasized the centrality of representations in learning and knowing science, the need to frame learning sequences
around the development of students' representational resources, the need to make
explicit the form and function of representations, and the need to develop meta-
representational competence. Further, we have drawn on a literature that goes further than emphasizing representational interpretation, to advocate representational
construction, negotiation and evaluation in authentic settings, in order to more
deeply engage students in the knowledge building practices of science.

Exploring the Role of Representation Construction in Learning Science

The major project that established the practical and theoretical principles of the
representation construction approach to guided inquiry was the *Role of
Representation in Learning Science* (RILS) project, which ran from 2007 to 2009.
RILS has spawned a series of subsequent projects, which will also be described.
Prior to that, members of the RiLS team had explored these principles in classroom
settings (Carolan et al. 2008; Waldrip et al. 2010) and become convinced of their
potential to engage students in quality learning. RiLS provided an opportunity to

explore more systematically the nature of the teaching and learning principles that might be built around representation construction, and the resulting quality of student engagement with learning. In addition to these classroom antecedents, close investigation of the generative role of representation work for students' grasp of key concepts, as part of a longitudinal study of students' reasoning and learning (Tytler et al. 2006, 2007; Prain et al. 2009; Tytler and Prain 2010), convinced us that representation work held an important key to the learning challenges identified in so much of the science education literature. We formed the view that conceptual change theory, while providing significant insights into the major task of moving from everyday to scientific perspectives, has not been up to now a theory of learning. The process by which students are inducted into scientific ways of perceiving the world is fundamentally representational, in that shifting from everyday to scientific conceptions involves the development of new representational languages. Thus, we have argued (Tytler and Prain 2010, 2013) that the challenge of conceptual change is fundamentally a representational challenge.

The RiLS project had a number of facets, but its central feature was an in depth exploration in which members of the team worked closely with two elementary, and two or sometimes three secondary teachers, to explore the approach applied to key science topics that were known to present learning challenges for students. The elementary school topics were animals in the school ground, energy, and water (changes to matter). The secondary school topics were force and motion, molecular models of substance, and astronomy. The teaching and learning approach involved constructing learning sequences with the teachers around a series of representational challenges that foregrounded assessment of representational adequacy and negotiation, and explicit consideration of the role of representations in learning and knowing. We worked with teachers across the middle years (5–9) of schooling, which pose particular difficulties for student engagement and where interest in science has been shown to markedly decline (Tytler and Osborne 2012). The pedagogy is consistent with what research would indicate is sound practice for teaching in these middle years, involving active engagement and challenge in learning activities, and higher order thinking and reasoning. The aim of the research was to:

(a) Systematically develop over these 3 years a set of principles of teaching and learning that exemplified our 'representation construction' position,
(b) Develop practical approaches that exemplify these principles,
(c) Investigate the challenges for teachers in adopting this approach, and
(d) Identify the student learning gains associated with the approach.

For each unit of work, the teachers' practices, student-teacher interactions, and student activity and discussion were monitored using classroom video capture. This involved two cameras arranged to film the teacher, and a selected group of students for each lesson. Radio microphones were used for teachers and the student group. The video was captured on digital tape and coded using Studiocode software to identify 'quality teaching and learning moments' judged fruitful for subjecting to further analysis. These teaching and learning sequences were then selectively transcribed and subjected to interpretive analysis to identify the teaching and learning principles underpinning them, and for evidence of the ways in which the focus on representations supported reasoning and learning. Students were interviewed about their learning and their understandings of the relationship between representations and knowing, and teachers about their perceptions of the effectiveness of aspects of the sequence. Student workbooks were collected to provide a continuous record of representational work. We held a series of workshops in which teachers and researchers reflected on and discussed their observations and experiences.

In working with the teachers over 3 years, we developed a set of teaching and learning principles based on a process of continual analysis of the data, interacting with the theoretical ideas described above in which the research was grounded. These were available to teachers, and were the working principles we used to help teachers plan the lesson sequence. While the broad principles were in place early in the project, the refinement presented here represents a growing understanding of the key elements and their relative emphasis, the relation between the different principles, and the detailed nature of the teaching practice and the student learning arising from each principle. The approach we developed reflects a view of quality learning as induction into the epistemic practices of the science community, with student construction of scientific representations understood as a crucial strategy for acquiring an understanding of the literacies of science as well as their underpinning epistemologies and purposes.

A Representation Construction Pedagogy

The set of principles that underpin the representation construction approach to teaching and learning are described as:

1. *Teaching sequences are based on sequences of representational challenges:* Students construct representations to actively explore and make claims about phenomena.

    (a) *Clarifying the representational resources underpinning key concepts :* Teachers need to clearly identify big ideas, key concepts and their representations, at the planning stage of a topic in order to guide refinement of representational work.

    (b) *Establishing a representational need:* The sequence needs to involve explorations in which students identify the problematic nature of phenomena and the need for explanatory representation, before the introduction of the scientifically accepted forms.

    (c) *Coordinating/aligning student generated and canonical representations:* There needs to be interplay between teacher-introduced and student-constructed representations where students are challenged and supported to refine and extend and coordinate their understandings.

2. *Representations are explicitly discussed:* The teacher plays multiple roles, scaffolding the discussion to critique and support student representation construction in a shared classroom process. Students build their meta-representational competency through these discussions.

   (a) *The selective purpose of any representation:* Students need to understand that a number of representations are needed for working with multiple aspects of a concept.

   (b) *Group agreement on generative representations* : There needs to be a guided process whereby students critique representations to aim at a resolution.

   (c) *Form and function:* There needs to be an explicit focus on representational function and form, with timely clarification of parts and their purposes.

   (d) *The adequacy of representations:* There needs to be ongoing assessment (by teachers and students) of student representations as well as those representations introduced by the teacher.

3. *Meaningful learning* : Providing strong perceptual/experiential contexts and attending to student engagement and interests through choice of task and encouraging student agency.

   (a) *Perceptual context:* Activity sequences need to have a strong perceptual context (i.e., hands on, experiential) and allow constant two-way mapping between objects and representations.

   (b) *Engagement / agency:* Activity sequences need to focus on engaging students in learning that is personally meaningful and challenging, through affording agency and attending to students' interests, values and aesthetic preferences, and personal histories.

4. *Assessment through representations:* Formative and summative assessment needs to allow opportunities for students to generate and interpret representations. Students and teachers are involved in a continuous, embedded process of assessing the adequacy of representations, and their coordination, in explanatory accounts.

To illustrate the key elements of representation construction approach examples will be given from complete topic sequences in forces, astronomy, energy, geology, and ideas about matter taught to Year 7/8 students who were between 12 and 13 years of age, and topic sequences in animals in the schoolyard and water taught to Year 5/6 students who were between 10 and 11 years of age. Several teachers from three different schools participated in the teaching of these topics.

In enacting a representation construction approach emphasis needs to be placed during the planning stage on the determination of the key concepts that underpin the topics to be taught [*Principle 1a*]. These concepts are expressed as statements of understanding couched in language readily understood by the students. For example, 'Force is a push or pull of one object, the doer, onto another object, the receiver' [Year 7 Forces]; and, 'Objects like the Earth and Moon spin, or rotate, on an axis, and revolve, or orbit, other objects' [Year 8 Astronomy]. Given that a concept consists of a set of interlinked representations and practices the statements of understanding guide the teacher in the development of the representational resources and strategies to use in teaching each concept. The statements also guide the teacher in developing a set of representational experiences that provide students with a coherent link between the concepts. The following sections provide examples of how three topics, ideas about matter, forces and astronomy were introduced.

Introduction to the Topic of Ideas About Matter

In introducing the topic of ideas about matter the teacher had a focus on the key concept that substances have a variety of physical and chemical properties that scientists categorize in various ways. The students were grouped in pairs. Each pair was given a representational challenge that asked them to sort 12 items, represented as labelled images on a sheet of paper, in two different ways, and to represent their groupings in their workbooks. The 12 items were sunscreen, towel, sea water, sand, rubber flipper, drinking water, oxygen in a tank, helium inside balloons, smoke, bubble wrap, slime and an apple that were cut into separate pieces of paper by the students. The video record showed the pairs of students physically manipulating the pieces of paper, using them as a representational tool to explore different groupings, to argue both the merits of one grouping over another and the placement of items into specific groups.

While most pairs of students sorted the items according to a classification schema of states of matter there was variation in the second schema used. For example, Natural/Non-natural items; Things that could be broken: Easy/Medium/Hard; and, Flexibility of item: Elastic/Pliable/Brittle. There were groupings without a specific schema, for example, Beach/Party/Everywhere (containing a single item, oxygen in a tank); and, Liquids/Stuff you breathe in/Party stuff/ Squishy Stuff/ something you use after a shower. A key element of the representational challenge is the evaluation of the student-generated representations in terms of how well they address the given task. Apart from the discourse that the teacher had with each pair as they were constructing their representations, a class discussion ensued whereby selected pairs presented and discussed their representations with the whole class. One issue that was discussed related to the affordances of creating schemas that might accommodate new items that are introduced versus groupings that do not have a particular schema. Through this discussion the teacher was aligning the students to the scientific practice of creating classification schemas that sort a multitude of items [*Principle 1c*].

The other issue raised by the class discussion of the student generated representations was the manner in which one might represent schema where items were represented in multiple groups; a representational need was established [*Principle 1b*]. The teacher used the classification schema of states of matter to introduce two representational forms, a Venn diagram and a continuum. Several of the students knew about Venn diagram but for the other students the teacher needed to explicitly

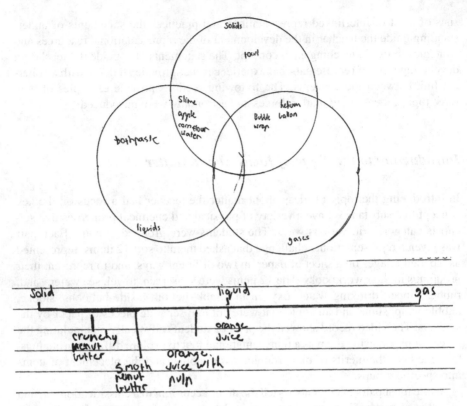

Fig. 9.1 Venn diagram and continuum representations of states of matter (Year 7)

discuss with students the function and form of Venn diagrams [**Principle 2c**]. This was also the case for the continuum representation which was new for all students. Figure 9.1 shows the class generated Venn diagram and continuum for various items generated by the teacher and students. These forms of representation were seen as important for the teacher as a significant amount of matter that students come into contact on an everyday basis is really a mixture of substances and therefore explorations of different states of matter needs to include situations where the 'stuff' that is being observed might be classified as a mixture of two states, represented in the Venn Diagram as the intersection of groups, or as a location somewhere on a continuum. The case of classifying bubble wrap, which many of the students argued was a mixture of a solid and a gas, was discussed in relation to the continuum representation. The teacher responded that this was a case where the representational form was restrictive in its ability to represent such an item as a mixture of a gas and solid [**Principle 2d**].

Introduction to the Topic of Astronomy

The teacher introduced the topic of Astronomy with a classroom discussion that involved critiquing one of the most universal canonical representations in astronomy, that of the globe as a representation of Earth in space. The critique involved the students generating lists of features of Earth that were represented by the globe as well as those features that were not represented by the globe [*Principle 2d*]. The task was given initially as an individual bookwork exercise followed by a classroom discussion. Table 9.1 gives examples of features of Earth generated by the students.

The teacher then used a globe with another object (ball) in the room that represented the Sun. A small figurine was affixed to the globe to represent an observer on Earth located in Melbourne, Australia. Rather than making explicit statements as to how this space view model might represent various observations that an observer on Earth might make the teacher provided a series of questions for the students to answer. For example:

1. Which way does the Earth spin?
2. For the position of the figurine on the globe what time of day is it? Where else on Earth is it the same time of day?

In answering the first question one student rotated the globe and reasoned that as the Sun shines on the East coast of Australia first then the Earth must rotate in an anti-clockwise fashion. Following this activity, and in subsequent lessons, each pair of students was given access to a mini-globe and a small, but powerful, Light Emitting Diode (LED) torch to be used as a reasoning tool to gain understanding of various astronomical behavior such as day and night cycle, eclipses, seasons and phases of the Moon. The teacher saw the use of the globe and torch as a tool by which the students might make links between a geocentric perspective or a phenomenon and its explanation that requires a space-centric perspective [*Principle 1a, c*].

The fundamental movements of rotation and revolution of celestial objects such as the Earth, Moon and Sun given rise to a variety of phenomena on Earth such as day and night cycle and phases of the Moon. The teacher gave the students a series of representational challenges that explored these movements. The first challenge was for a pair of students to demonstrate, through a role play, their understanding of rotation and revolution. Following a class discussion that evaluated the role plays for their representational adequacy all students came to an agreed understanding of

Table 9.1 Features of Earth represented/not represented by the globe

What does the globe represent?	What does the globe NOT represent?
The axis of the Earth is tilted	Clouds/atmosphere
Shape (round)	Gravity
The earth rotates (spins)	Day and night cycle
Continents and oceans	Size of Earth
	Inside the Earth

these movements and the need for a central axis for rotation and a central point for revolution and linkage of these terms with the everyday language of spin (rotate) and orbit (revolve).

The teacher then provided the students two challenging tasks in the form of representational challenges [*Principle 1*]. These were:

1. Is it possible for two celestial objects to revolve about each other?
2. The Moon always has one face to the Earth. Over 1 month the Moon undertakes one complete revolution of the Earth. During this time does it also rotate and, if so, how many times?

The students found that answers to these challenges could only be obtained through role-play [*Principle 1b*]. The evaluation phase of the challenges were undertaken as a whole class discussion. For the first challenge, which the answer is yes, the teacher made a link to binary star systems where this phenomenon is found. For the second challenge, many of the students were quite skeptical in the beginning as to whether the Moon rotated but by undertaking the role play they found evidence that the Moon does indeed rotate, making one full rotation each month. This evidence was the observation that in undertaking one full revolution the student representing the Moon observed each wall of the classroom just once. Figure 9.2a shows evidence of a third challenge given to the students whereby they were to pictorially represent two objects revolving about each other. The teacher had a common practice that in most lessons some time was allocated at the end of the lesson for the students to represent in their journals something they had learned that lesson. Figure 9.2b shows one student's representation of the motion of the Moon.

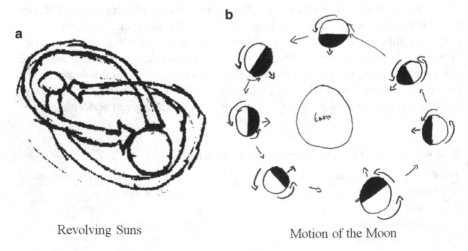

Revolving Suns   Motion of the Moon

Fig. 9.2 Year 8 representations of revolving suns and motion of the moon

Evidence of Student Learning

Pre- and post-testing of students in classrooms where a representational construction approach was adopted have indicated substantive learning gains. While there was anecdotal claims from the participating teachers as to an enhanced engagement and learning by the teaching approach the initial research project described here did not apply an experimental design to support such claims. However, we were able to make comparisons of the pre- and post-testing of core astronomy concepts in two of our research schools and make a comparison with a separate international study (Kalkan and Kiroglu 2007) who used the same instrument.

Table 9.2 indicates the students' results of responses to 13 multiple choice questions on the pre/post-tests for two schools and provides a comparison to results obtained by the Kalkan and Kiroglu (2007) study. The first school in our research involved two teachers and two Year 8 classes of students in 2008 while the second school involved four teachers and five Year 8 classes of students in 2012. The Kalkan and Kiroglu (2007) study involved 100 pre-service primary and secondary education teachers who participated in a semester length course in astronomy. A measure of comparison of pre- and post-test results is the normalized gain index, <g>, the ratio of the actual average student gain to the maximum possible average gain: <g>= (post% − pre%)/(100 − pre%), reported by Zeilik et al. (1998). Gain index values can range from 0 (no gain achieved) to 1 (all possible gain achieved).

Table 9.2 Correct answer ratio and gain index (<g>) according to pre- and post-test results for three studies

Item	Year 8 (2008) N = 33			Year 8 (2012) N = 125			Kalkan and Kiroglu (2007) study N = 100		
	Pre-test	Post-test	Gain	Pre-test	Post-test	Gain	Pre-test	Post-test	Gain
Day-night cycle	9	91	0.80	61	92	0.78	91	93	0.22
Moon phases	30	55	0.36	43	81	0.66	23	30	0.09
Sun Earth distance scale	19	60	0.44	9	49	0.44	18	22	0.05
Altitude of midday Sun	8	56	0.53	10	66	0.62	29	39	0.14
Earth dimensions	12	57	0.44	30	63	0.48	5	14	0.09
Seasons	8	28	0.23	13	63	0.57	54	82	0.61
Relative distances	33	62	0.49	70	85	0.51	46	71	0.46
Moon's revolution	27	82	0.72	39	83	0.72	49	60	0.22
Sun's revolution	60	85	0.70	86	97	0.79	61	77	0.41
Solar eclipse	18	41	0.32	31	86	0.79	26	42	0.22
Moon's rotation	15	55	0.48	21	61	0.5	13	28	0.17
Centre of universe location	61	76	0.48	78	95	0.75	65	88	0.66
Seasons	38	89	0.81	73	97	0.89	67	88	0.64
	Mean <g>		0.65	Mean <g>		0.52	Mean <g>		0.31

The mean gain reported by Kalkan and Kiroglu (2007, p. 17) was described as a "respectable 0.3". In contrast the mean gain for the two schools in our research was significantly higher at 0.52 and 0.63.

Meaningful Learning Through Representation

The third principle of the representation construction approach makes reference to the need to provide strong perceptual/experiential contexts for the students. In a Year 5/6 'Animals in the school ground' sequence students, after mapping the incidence and diversity of invertebrates in different locations, were challenged to represent a particular animals' movement through a 3D model, first making design drawings. The challenge stimulated children to examine closely and discuss details of the animal's movements, abstracting essential elements in their representations (*Principle 3a*). The teachers scaffolded this process of attention, selection and abstraction by challenges such as: "You don't have to draw the exact creature, but you have to show how it actually moves. [Consider the] order, when that one goes when, [points] … when that one goes when." "Now do they move together? At the same time? So how are you going to show me that?" Figure 9.3 shows an annotated drawing of an earthworm's movement, and the abstracted construction through which Ivan and his partner wanted to model as accurately as possible the amount of extension and retraction of the earthworm. To this end they drew up a scale on an A3 paper to help them represent the exact extension as the earthworm moved along

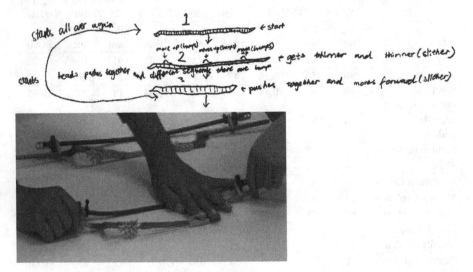

Fig. 9.3 Ivan and his partner's drawing and modeling of earthworm movement

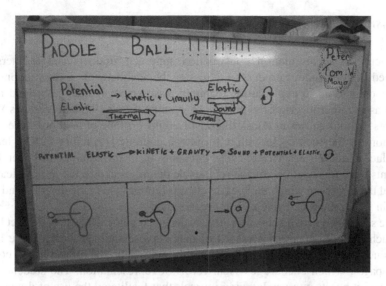

Fig. 9.4 Year 8 group's representation of the operation of a child's toy and energy changes in operation of the toy

a smooth surface. The model was constructed of meccano pieces, connects, flexi wire and blu-tac. Ivan describes their investigation of the model:

> So we tested it out by using blu-tac and a rubber band to see which one would represent better. So what we did with the blu-tac… we basically pull it one way [pulls the mechanical handle] this is how it moves. It moves forward one way, it goes a bit at the back … then keeps repeating it…. And then we tried doing the same with the rubber band and the rubber band worked well because it retracted really easily.

In the topic of Energy the students were placed in groups of four and provided with a child's toy. Their challenge was to explain how the toy operated and the energy changes that take place in the operation of the toy. Their explanations were to be initially shown on a mini-whiteboard and supplemented by a verbal presentation. Figure 9.4 shows one group's whiteboard representation. They used a time sequence of drawings with arrows to represent the operation of the toy and employed the canonical representation forms of a Sankey diagram and energy flow diagram to represent the energy changes. In completing the task the students made regular transitions between manipulating the toy and making inscriptions on the whiteboard [***Principle 3a, b***]. The mini-whiteboard allowed the students to readily edit their inscriptions but constrained them in what they could represent given the limited space. The students had an emphasis on diagrammatic forms of representation which they were able to supplement with their verbal presentation. The adequacy of their representations in relation to the set task was discussed in class.

Assessment Through Representations

In enacting a representation construction approach students and teachers are involved in a continuous process of assessing the adequacy of representations, and their coordination, in explanatory accounts. Assessment is thus embedded in the lesson as an integral part of the process of learning. Figure 9.5 below shows Year 8 students' journal entries of a task that involved the critique of textbook-type representations of the rock cycle within a topic on geology [*Principle 4*]. The teacher reproduced four different rock-cycle diagrams from the internet in paper form. Students were to select one of the rock-cycle diagrams and to list in their learning journal three ways in which the rock-cycle was represented in the diagram and three ways in which the rock-cycle was not represented.

One school we worked closely with developed an innovative use of project books in which students maintain a record of their representational work. These books have since become an important feature of the approach. The project books contain blank pages, which encourage visual forms of representation. The students used their project books more as learning journals that facilitated the use of drawings in recording not only what they had learned but their developing ideas (see Fig. 9.6 for examples). The visual representations provided the teacher with ready insight into students' thinking. One of the teachers, Alice, commented in an interview:

> Immediately by looking at their representations, I know, okay those boys have got it and those boys are on the right track but those haven't fully kind of understood (Alice)
>
> But the books just having the blank page, I think sometimes, it's just all text, that we kind of forget how much the use of those representations and diagrams can really help in Science, so it was a good reminder. (Alice)

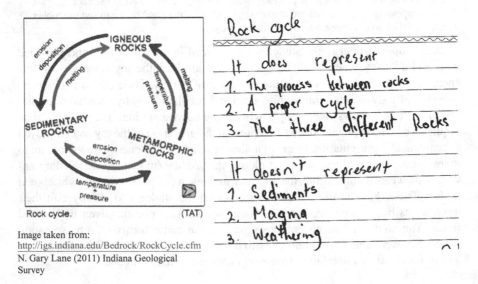

Image taken from:
http://igs.indiana.edu/Bedrock/RockCycle.cfm
N. Gary Lane (2011) Indiana Geological
Survey

Fig. 9.5 Two Year 8 students' critique of a representation of the rock-cycle found on the internet

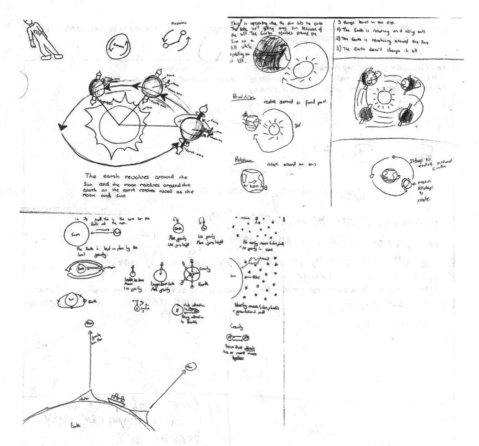

Fig. 9.6 Examples of students' learning journals entries

The teachers found the students were more willing to use their journals to reflect on their learning. As Kate and Alice pointed out:

> …they seemed more willing to go back over their work and look back at their past stuff as well…And I don't think they do it very well if it's just written stuff and they had a sense of ownership over it which was good (Kate)
>
> They loved their project books. Like ridiculously…. it was like this little diary of all the work that they'd done. It was different from what they had been doing. (Alice)

Teachers have also found that the ample provision of space to respond to questions on paper-based tests affords the students the opportunity to express their understanding using combinations of text and multiple representational forms. For example, Fig. 9.7 shows a Year 6 students' response to the question: "Where do you think the water in the tiny droplets of water in the clouds comes from?" And "Use representations to show how little drops of water form clouds". This post-test followed a unit on evaporation where students had been exposed to particle ideas for evaporative phenomena, but not condensation. The space allowed this student to construct a detailed narrative of the condensation process.

3. Use words and other representations (e.g. drawings, graphs, arrows) to represent what is happening in the stories below.

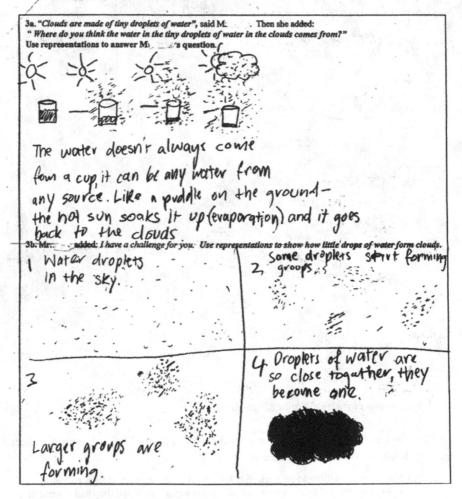

3a. *"Clouds are made of tiny droplets of water"*, said M. . Then she added:
" Where do you think the water in the tiny droplets of water in the clouds comes from?"
Use representations to answer Mi 's question.

The water doesn't always come
from a cup, it can be any water from
any source. Like a puddle on the ground—
the hot sun soaks it up (evaporation) and it goes
back to the clouds

3b. Mr: added: *I have a challenge for you. Use representations to show how little drops of water form clouds.*

1 Water droplets in the sky.

2 Some droplets start forming groups.

3 Larger groups are forming.

4 Droplets of water are so close together, they become one.

Fig. 9.7 A Year 6 student's representation of how little drops of water form clouds

Extending the Research into Teacher Development and Digital Pedagogy

In 2010–2012 the team received funding from the Victorian Department of Education and Training (DET) to develop and run intensive workshops for secondary science teachers across the state. This program, Switched on Secondary Science Professional Learning (SOSSPL), was based on the guided inquiry representation construction approach and involved teachers developing, implementing and reporting on short representational challenge sequences. Analysis of these reports and interviews with the teachers showed that teachers found the approach readily understood and attractive, and engaging for students (Hubber et al. 2011). We have no evidence however that they reached that deeper understanding of the epistemological basis of the approach that our RiLS teachers developed over time. In addition, the team has been involved, with other colleagues, in designing and delivering an extended professional learning program for specialist teachers of elementary science, again an initiative of the DET. The workshops again have been built around the representation construction approach to teaching and learning, and again teachers have found the approach immediately understandable and appealing, as an approach that engages students in active and imaginative processes.

In 2012 we began a research program investigating the processes by which the approach can be implemented more widely in schools. The CRISP project (*Constructing Representations in Science Pedagogy*) involves working with schools to investigate the nature of support needed to implement the approach in schools, and the processes by which teacher learning occurs as they engage with the ideas and pedagogy. To date, the research has identified a number of challenges, and successes, for teachers, and the diverse professional learning pathways they take in implementing the approach. Some of these issues are described in the vignettes that follow.

The Nature of Representational Challenges

In the RILS project we had worked closely with teachers to develop units of study with representational challenge sequences deliberately structured to cohere around key concepts. However, without the close involvement of the researchers, some teachers in CRISP have preferred to treat the challenges as staged singular events – as engaging activities that motivate students and engage them with learning but which do not necessarily build concepts in a deliberate way. This shifted the focus of attention, for some schools, to the representational challenge as carrying the core burden of change rather than the sequencing based on analysis of the core conceptual structure of topics. This also led to some confusion about the key nature of representational challenges. There was a need to articulate more clearly the distinction between the types of challenge at the centre of the approach which require students to go beyond what they are given and construct representations that speak to the agreed problem, and are in some sense fresh and unrehearsed, compared to the challenges commonly found in textbooks that require application of known representations. An example of the latter would include problems where a representation of the earth-moon-sun system is given and students are required to use it to answer a series of questions such as which points on earth are in daylight, or where the moon is full.

Working with these teachers also forced us to examine the importance of appropriate scaffolding in constructing an appropriate challenge to ensure students' representational moves are productive. Looking through the examples of challenges in the RILS book (Tytler et al. 2013) makes it immediately apparent that the challenges are extremely varied and that their generation requires some experience and skill. We are currently working on the project website to produce a battery of sequences of representational challenges, illustrative of the approach. A distinction that became clear during these conversations as a part of the CRISP program is that between (a) setting a challenge 'from the ground up' where students are not given models to work by, such as the elementary school class that discussed how to compare animal populations in different ecological niches in the school ground and were gradually led to a version of quadrats as a comparative device, compared to (b) more abstract and sophisticated models such as that of the earth-moon-sun system where the model is produced and the challenges are representational re-descriptions involving changes in perspective, in dimensionality, or in mode. We had found that the teachers we worked with in RILS became confident in setting challenges as an integral part of their teaching, and made appropriate decisions about the context of the challenge and the degree of support, linking challenges in the way an experienced teacher can put together sequences of questioning that expand and refine understandings. Clearly, this expertise is hard won and requires substantial support to achieve. It involves an epistemological shift, and also the development of fluency in managing classroom interactions. The following vignette, of a school to which the team offered substantial support, will illustrate the processes by teachers can achieve such shifts.

The Professional Growth of Teachers in Developing Expertise

As part of the CRISP project the research team initially worked with a group of four teachers from a metropolitan all-boys Catholic secondary school who taught the topic of astronomy to their Year 8 classes. The teachers were supported with curriculum resources and professional development given to the teachers in various forms by the CRISP research team. Each of the teachers was given curriculum resources that were based on the findings from the teaching of this topic using the representation construction approach in one of the schools who took part in the RILS project. These resources consisted of a pre- and post-tests (with results from the RILS project), written descriptions of various activities that illustrated the representational approach, examples of students work from RILS project and digital resources in the form of PowerPoint presentations with embedded interactive simulations and video. The teachers participated in an initial 2-hour workshop conducted by the CRISP researchers at the school. Support was also given in weekly meetings that were conducted during the teaching sequence where CRISP researchers and teachers had reflective discussions as to the previous week's teaching in addition to planning the future week's teaching.

It was found that the teachers enacted many elements of the representation construction approach in the teaching sequence. They modified some of the resource materials in addition to introducing others as they tried different approaches and reflected on their teaching. Key changes to their normal practice included using a conceptual focus to plan topics, the pre-testing of students, the use of project books and decreased reliance on text books, and representational work including the implementation of conceptual challenges, the explicit discussion of representations, and more focus on multiple representations in summative assessment. Several of these changes were associated with students' use of project books that replaced their lined workbooks, as described earlier. The collaboration among the teachers in the past was "*more where are you up to in the text book rather than what the activities are* (Jaz)". The experience in teaching the Astronomy topic was more collaborative in the sense that there was a joint commitment to supporting deeper level student understanding around a conceptual focus. It was more about discussing and sharing representational issues with teaching and learning the concepts.

> I think we kind of went away from what's in the chapter of the text book that we need a cover into more what key ideas or understanding that we want the boys to take away at the end of the topic. (Alice)

Rather than following the textbook, the introduction of a pedagogy built around a coherent conceptual focus gave the teachers, "*a bit more fluidity as well and flexibility with what we can do as we go (Alice)*." With the collaborative planning of topics from a conceptual focus there is now an emphasis on pre-testing of ideas (mentioned above) and, "*being very conscious of trying to put in representation stuff (Kate)*".

Following the astronomy topic the teachers maintained their diminished focus on the textbook. This finding is reflected in the following comment:

> I'm not as text book oriented. I found that I don't set questions out of text book or the E book anymore I just give them task that suit where they're at and there will be homework tasks that follow on, it's more of a representational deal (Jaz).

The affordances offered by the representation construction pedagogy for collaboration around student conceptual learning also provided the opportunity for mentoring work amongst the teachers, with those with more experience or productive ideas helping the others, especially new staff as they came on board, with understanding and implementing the approach in novel ways.

Current Research Directions

An increasing number of Australian schools are moving to 1-1 situations where each student has a laptop or tablet device, and this has created a challenge to develop effective digital pedagogies. One school involved with CRISP operates their program in a cloud based environment, in which students negotiate a set of 'task cards' that provide web based information and activities, set practical challenges to be done during the school day, and specify assessment activities that must be completed and submitted online. In this 'flipped' classroom model, teachers are on hand during the school day to help student apply and negotiate knowledge they have accessed online. While teachers were impressed at the sophistication of some students' representational work, it became clear that two important ingredients were missing. These were the lack of guided practical experience to link to representational work, and particularly the lack of opportunity for student representations to be challenged and negotiated by teachers or peers. Currently the school is working to introduce more of these features into their approach. This need to introduce more inter-personal interactions and hands on activities into digitally-enabled classrooms is the focus of a current project which aims at developing an inquiry based digital pedagogy utilizing digitally produced representations. We are working with cloud-based planning and delivery software, STILE, which allows the setting of tasks across a lesson and students uploading responses including constructed representations, for display and discussion. This design-based research is working iteratively with schools to generate sequences of experimental activities and representational/ modeling challenges that can be negotiated online and face-to-face in classrooms. Our hope is to develop the software, and an approach to combining digital interaction, experimental work, and classroom discussion, to create an effective approach to inquiry learning in 1-1 digital classroom environments.

A second project is part of a large 'Science of Learning Research Centre' dedicated to aligning classroom perspectives on learning with neuroscience findings. The particular projects involving representation construction will use intensive video capture and micro-ethnographic analysis to focus on the affordances for learning offered by dyads and groups of students engaged in constructing representations. Our longer-term plan is to search for theoretical and methodological bridges that can span the all too obvious gap between socio cultural perspectives of the science classroom, and reductive accounts of learning that by necessity dominate the neuroscience literature.

Concluding Comments

This chapter has described a substantial program of research over a decade that has explored representation construction as a critical feature of quality learning in science. The journey has involved a range of researchers, and has been informed by increasingly confident theoretical work built around socio-cultural and pragmatist perspectives on learning and knowing. The research has yielded compelling evidence of the learning advantages conferred by the approach, and its appeal to teachers and policy makers as a productive way forward. It has also highlighted the epistemological and pedagogical challenges for teachers in developing the approach, but encouraging examples of where professional growth has occurred. As our ideas are accessed by more teachers and by science and education agencies, and as we build more resources to support teachers, we are hopeful that over time this representation construction approach to guided inquiry will become a significant aspect of how quality teaching and learning in science is conceived of in Australia.

Acknowledgements We would like to acknowledge the many academics who have been part of the teams working on this approach over a number of projects: Vaughan Prain (La Trobe University), Bruce Waldrip (University of Tasmania), Gail Chittleborough and George Aranda (Deakin University), Peter Aubusson (University of Technology, Sydney), and Garry Hoban (University of Wollongong). We also acknowledge the Australian Research Council for their funding of the RILS, CRISP, iSTELR and SLRC projects.

References

Ainsworth, S. (2006). DEFT: A conceptual framework for learning with multiple representations. *Learning and Instruction, 16*(3), 183–198.

Ainsworth, S. (2008). The educational value of multiple representations when learning complex scientific concepts. In J. K. Gilbert, M. Reiner, & M. Nakhlel (Eds.), *Visualization: Theory and practice in science education* (pp. 191–208). New York: Springer.

Ainsworth, S., Prain, V., & Tytler, R. (2011). Drawing to learn in science. *Science, 333*, 1096–1097.

Bransford, J., & Schwartz, D. (1999). Rethinking transfer: A simple proposal with multiple implications. *Review of Research in Education, 24*, 61–100.

Carolan, J., Prain, V., & Waldrip, B. (2008). Using representations for teaching and learning in science. *Teaching Science, 54*, 18–23.

Cazden, C. (1981). Performance before competence: Assistance to child discourse in the zone of proximal development. *The Quarterly Newsletter of the Laboratory of Comparative Human Cognition, 3*, 5–8.

Clement, J. (2000). Model based learning as key research area for science education. *International Journal of Science Education, 22*(9), 1041–1053.

Clement, J., & Rea-Ramirez, M. (Eds.). (2008). *Model based learning and instruction in science*. Dordrecht: Springer.

Cole, M., & Wertsch, J. V. (1996). Beyond the individual-social antinomy in discussions of Piaget and Vygotsky. *Human Development, 39*, 250–256.

Cox, R. (1999). Representation construction, externalized cognition and individual differences. *Learning and Instruction, 9*, 343–363.

diSessa, A. (2004). Metarepresentation: Native competence and targets for instruction. *Cognition and Instruction, 22*(3), 293–331.

Ford, M., & Forman, E. A. (2006). Refining disciplinary learning in classroom contexts. *Review of Research in Education, 30*, 1–33.

Gibson, J. (1979). *The ecological approach to visual perception*. Boston: Houghton Mifflin.

Gilbert, J. K. (2005). *Visualization in science education*. New York: Springer.

Gooding, D. (2004). Visualization, inference and explanation in the sciences. In G. Malcolm (Ed.), *Studies in multidisciplinarity* (Vol. 2, pp. 1–25). Amsterdam: Elsevier.

Gooding, D. (2006). From phenomenology to field theory: Faraday's visual reasoning. *Perspectives on Science, 14*(1), 40–65.

Greeno, J. G. (2009). A theory bite on contextualizing, framing, and positioning: A companion to Son and Goldstone. *Cognition and Instruction, 27*(3), 269–275.

Greeno, J. G., & Hall, R. P. (1997). Practicing representation: Learning with and about representational forms. *Phi Delta Kappan, 78*(5), 361–368.

Hubber, P., Tytler, R., Chittleborough, G., & Campbell, C. (2011). *Report on the switched on secondary science professional learning program*. Burwood: Deakin University.

Kalkan, H., & Kiroglu, K. (2007). Science and nonscience students' ideas about basic astronomy concepts in preservice training for elementary school teachers. *Astronomy Education Review, 1*(6), 15–24.

Klein, P. D. (2000). Elementary students' strategies for writing-to-learn in science. *Cognition and Instruction, 18*(3), 317–348.

Klein, P. (2006). The challenges of scientific literacy: From the viewpoint of second-generation cognitive science. *International Journal of Science Education, 28*(2–3), 143–178.

Kozma, R., & Russell, J. (1997). Multimedia and understanding: Expert and novice responses to different representations of chemical phenomena. *Journal of Research in Science Teaching, 34*(9), 949–968.

Kozma, R. B., Chin, E., Russell, J., & Marx, N. (2000). The roles of representations and tools in the chemistry laboratory and their implications for chemistry instruction. *Journal of the Learning Sciences, 9*(2), 105–143.

Kozma, R., & Russell, J. (2005). Students becoming chemists: Developing representational competence. In J. Gilbert (Ed.), *Visualization in science education* (pp. 121–146). London: Kluwer.

Latour, B. (1999). *Pandora's hope: Essays on the reality of science studies*. Cambridge, MA: Harvard University Press.

Lave, J., & Wenger, E. (1991). *Situated learning: Legitimate peripheral participation*. Cambridge/New York: Cambridge University Press.

Lehrer, R., & Schauble, L. (2006). Cultivating model-based reasoning in science education. In K. Sawyer (Ed.), *Cambridge handbook of the learning sciences* (pp. 371–388). Cambridge, MA: Cambridge University Press.

Lemke, J. (2004). The literacies of science. In E. W. Saul (Ed.), *Crossing borders in literacy and science instruction: Perspectives on theory and practice* (pp. 33–47). Newark: International Reading Association/National Science Teachers Association.

Manz, E. (2012, April). *Engaging students in the epistemic functions of scientific argumentation*. Paper presented at the annual conference of the American Educational Research Association, Vancouver.

Moje, E. (2007). Developing socially just subject-matter instruction: A review of the literature on disciplinary literacy learning. *Review of Research in Education, 31*, 1–44.

Nersessian, N. (2008). Model-based reasoning in scientific practice. In R. Duschl & R. Grandy (Eds.), *Teaching scientific inquiry: Recommendations for research and implementation* (pp. 57–79). Rotterdam: Sense Publishers.

Peirce, C. S. (1931–1958). *Collected papers of Charles Sanders Peirce*. 8 Volumes (Eds., C. Hartshorne, P. Weiss, & A. W. Burks, Vol. 1–6), (Ed., A. W. Burks, Vol. 7–8). Cambridge, MA: Harvard University Press.

Pickering, A. (1995). *The mangle of practice: Time, agency and science*. Chicago: University of Chicago Press.

Prain, V., & Tytler, R. (2012). Learning through constructing representations in science: A framework of representational construction affordances. *International Journal of Science Education, 34*(17), 2751–2773.

Prain, V., & Tytler, R. (2013). Learning through the affordances of representation construction. In R. Tytler, V. Prain, P. Hubber, & B. Waldrip (Eds.), *Constructing representations to learn in science* (pp. 67–82). Rotterdam: Sense Publishers.

Prain, V., Tytler, R., & Peterson, S. (2009). Multiple representation in learning about evaporation. *International Journal of Science Education, 31*(6), 787–808.

Tytler, R., & Osborne, J. (2012). Student attitudes and aspirations towards science. In B. Fraser, K. Tobin, & C. McRobbie (Eds.), *Second international handbook of science education* (pp. 597–625). Dordrecht: Springer.

Tytler, R., & Prain, V. (2010). A framework for re-thinking learning in science from recent cognitive science perspectives. *International Journal of Science Education, 32*(15), 2055–2078.

Tytler, R., & Prain, V. (2013). Representation construction to support conceptual change. In S. Vosniadou (Ed.), *Handbook of research on conceptual change* (pp. 560–579). New York: Routledge.

Tytler, R., Peterson, S., & Prain, V. (2006). Picturing evaporation: Learning science literacy through a particle representation. *Teaching Science, the Journal of the Australian Science Teachers Association, 52*(1), 12–17.

Tytler, R., Prain, V., & Peterson, S. (2007). Representational issues in students learning about evaporation. *Research in Science Education, 37*(3), 313–331.

Tytler, R., Haslam, F., Prain, V., & Hubber, P. (2009). An explicit representational focus for teaching and learning about animals in the environment. *Teaching Science, 55*(4), 21–27.

Tytler, R., Prain, V., Hubber, P., & Waldrip, B. (Eds.). (2013). *Constructing representations to learn in science*. Rotterdam: Sense Publishers.

Vosniadou, S. (1994). Capturing and modeling the process of conceptual change. *Learning and Instruction, 4,* 45–69.

Vygotsky, L. (1981). *Thought and language* (revised and edited by A. Kozulin). Cambridge, MA: MIT Press.

Waldrip, B., Prain, V., & Carolan, J. (2010). Using multi-modal representations to improve learning in junior secondary science. *Research in Science Education, 40*(1), 65–80.

Wittgenstein, L. (1972). In G. E. M. Anscombe, & R. Rhees (Eds.), *Philosophical investigations* (trans: Anscombe, G. E. M.). (2nd ed.). Oxford: Basil Blackwell.

Zeilik, M., Schau, C., & Mattern, N. (1998). Misconceptions and their change in university-level astronomy courses. *The Physics Teacher, 36*(1), 104–107.

Zhang, J. (1997). The nature of external representations in problem solving. *Cognitive Science, 21*(2), 179–217.

Chapter 10
Modeling Scientific Communication with Multimodal Writing Tasks: Impact on Students at Different Grade Levels

Mark A. McDermott and Brian Hand

Introduction

Emerging definitions of science literacy emphasize the dynamic interaction between conceptual aspects of understanding science and the literacy skills related to this understanding (Yore 2012). These increasingly expansive definitions of science literacy call for, among other things, development in students of appropriate cognitive and metacognitive practices associated with learning science, helping students utilize reasoning strategies more appropriately to improve their critical thinking and develop habits of mind, and naturally infusing "scientific language arts" into the curriculum through a focus on reading, writing, speaking, listening, viewing, and representing science. At the same time, it has been suggested science literacy must also involve the generation of big ideas and unifying conceptual understanding and engagement in and modeling of the nature of science as students learn with exploration of science through inquiry (Yore 2012). Yore (2012) summarizes a position promoting these varied aspects of science literacy in what he calls "Vision III – Science Literacy for All".

The curricular interventions advocated as a means of accomplishing the goal of helping students achieve this type of science literacy are as varied and as multifaceted as the emerging definition itself. As science educators develop potentially beneficial learning practices aligned with this definition of science literacy, they must attempt to clarify the relationship between the cognitive demand students encounter and the impact on the development of science literacy for these students. In this

M.A. McDermott (✉)
College of Education, N236 Lindquist Center, University of Iowa, Iowa City, IA 52242, USA
e-mail: mark-a-mcdermott@uiowa.edu

B. Hand
University of Iowa, Iowa City, IA, USA
e-mail: brian-hand@uiowa.edu

© Springer International Publishing Switzerland 2016
B. Hand et al. (eds.), *Using Multimodal Representations to Support Learning in the Science Classroom*, DOI 10.1007/978-3-319-16450-2_10

article, one promising intervention, the use of multimodal writing-to-learn tasks will be explored through the lens of Cognitive Load Theory (CLT).

Literature Review

Historically, two major goals have existed in relation to the issue of student development of science literacy. First many authors have attempted to clarify exactly what is necessary for students to attain in order to be considered scientifically literate (National Research Council 1996; Hand 2007; Yore 2012). The scope of this paper does not include delving into this admittedly important task. Rather, the authors take the position that while the issue is not settled, the multifaceted "Vision III" of science literacy mentioned previously provides an appropriate definition which aligns generally with present standards documents as well as research findings related to best practices. The second major goal associated with science literacy, which directly relates to the intent of this paper, is to determine curricular interventions and practices that are most likely to improve student opportunities for gaining this multi-faceted science literacy. Beginning to address this issue, Yore (2012) states that pedagogy related to his Vision III would emphasize:

> science as inquiry, argument and constructing knowledge claims and explanations of patterns in nature and naturally occurring events; the essential constructive, persuasive, and communicative functions of languages in doing science; the unique conventions, traditions, and metalanguage in scientific discourse; and the functionality of language in teaching and learning science. (Yore 2012, p. 3)

Writing-to-Learn

Cognitive models describing potential benefits for students when writing tasks are set up in specific ways have been offered for at least the past 40 years (Emig 1977; Bereiter and Scardamalia 1987; Galbraith 1999; Klein 2006). In general, these models posit that when writing tasks are constructed with specific characteristics, it is possible to move beyond writing by students filling a purely communicative function toward a situation in which the writing task itself initiates further learning for the students. Bereiter and Scardamalia (1987) describe this as a transition from "knowledge-telling" to "knowledge transformation" and Galbraith (1999) even suggests writing can promote "knowledge generation". All these models, while using different terminology, ascribe to the general notion that writing tasks can provide an opportunity in which rhetorical constraints, techniques, and skill interact with current conceptual understanding in a way that allows for student consideration of what is lacking in the current understanding, student drive to acquire a better or fuller understanding, and then student communication of this improved understanding (McDermott and Hand 2010a).

This beneficial cognitive action, however, is not a guaranteed result of simply assigning a writing task about science (Klein 2006). Multiple studies support the contention that writing tasks must be set up in specific ways to encourage beneficial student cognition. Prain (2006) indicates that if the goal for students is to move beyond simple recall toward the ability to strengthen connections and gain deep understanding, then the writing tasks must provide opportunity for students to "paraphrase, re-word, elaborate, unpack and re-represent meanings, express uncertainties, analyze comparisons, and reconstruct understandings" (p. 185). Prain and Hand (1996) argue that in order to create these beneficial activities, teachers must move away from solely utilizing traditional writing styles (formal lab write ups and written responses on worksheets) and embrace alternative writing tasks. In these alternative writing tasks, teachers can manipulate the type of writing (narrative, brochures, powerpoints), the audience written to (peers, younger students, parents), the way the text is produced (handwritten, computer generated), and the goal of the writing (summarizing, exploration, clarification) as they develop student activities to deal with multiple science concepts. Manipulation of these factors in the creation of writing tasks has been shown to be beneficial for students (Gunel et al. 2009) and students themselves have indicated a recognition of greater conceptual understanding when engaged in these activities (McDermott and Hand 2010b). One ongoing area of research involves the infusion of multimodal representations into these alternative writing tasks where students are asked to combine multiple ways of representing science conceptual information (diagrams, charts, graphs, equations) with text.

Multimodal Representations

The use of multiple modes to represent information and ideas is widespread in the discipline of science. A brief look at any science textbook, article in a science periodical, or science research poster would provide evidence that scientists use a broad array of modal representations (pictures, diagrams, charts, tables, equations) in communicating about scientific knowledge. Klein (2001) argues that the use of multimodal paper tools like diagrams, charts, and models was crucial in moving science forward historically. The ability to represent science understanding in multimodal ways allowed for greater communication about science research and the ability to improve and build science knowledge (Klein 2001). This historical progression highlights several issues relevant to science education today. Building scientific understanding among students necessitates the development of multiple literacies as a means of dealing with the multiple ways science understanding is displayed (Alvermann 2004). In addition, developing strategies to appropriately communicate science understanding in a multimodal way would seem to be an important skill in learning to communicate like a scientist. The ultimate goal, as Seufert (2003) points out, is to help students develop connections both between and within multiple representations of science ideas. In doing so, students will attain a deeper understanding.

The types of representations used in science can be broadly categorized into two main groups. Dorfner (1999) suggests that some representations mirror an actual structure or substance in reality. A diagram of the structure of an organic molecule would be this type of representation. A second type of representation would include representations like the periodic table of the elements. These representations are tools which "do not directly correspond to anything in the world", but through a "complex individual interaction of an agent with its environment" can support the development of new knowledge (Dorfner 1999, p. 25). This emphasis on the interaction between an individual and different modes of representation reiterates the idea previously mentioned in describing the potential of writing-to-learn activities of not only providing a vehicle for communication of already understood material, but of also providing a means for initiating the development of transformed or newly generated understanding. Kress et al. (2006) promote the idea of harnessing multimodal usage as a way to encourage meaning-making in science.

Theoretically, then, engagement with multimodal activities in the science classroom in which students are asked to produce alternative writing products including additional modes of representation beyond text should provide a potential route for student conceptual understanding development. Waldrip et al. (2010) suggest that involving students in activities in which they are required to "translate" information from the "language" of one mode to another encourages learning. However, since most students do not generally experience a great deal of multimodal writing during their educational progression, assigning these tasks will likely increase the cognitive complexity of the classroom action. In essence, students would be dealing cognitively with at least two main issues as they approach science concepts through multimodal writing tasks. First, students will be developing an understanding of a science concept that they are to some degree unfamiliar with. In addition, students will be engaging in a literacy task that they may not have developed sufficient skill to accomplish. Therefore, a major instructional question becomes whether this additional cognitive demand placed on the students is beneficial. Exploration of Cognitive Load Theory (CLT) provides a framework for discussing this issue.

Cognitive Load Theory (CLT)

CLT has been promoted both as an appropriate way to conceptualize cognition in humans and as a way to improve teaching and learning based on understanding this view. The basic premise of CLT is "that in order to be effective, instruction has to be adapted to the structure and functioning of the learners' cognitive architecture" (Schnotz and Kürschner 2007, p. 474). Several basic assumptions frame this theory and the understanding of cognitive architecture that it is based upon (Schnotz and Kürschner 2007). Each assumption is summarized below:

1. *Multiple Memory Stores*: CLT describes working memory as the relatively temporary, limited memory store in which humans process information they become sensory aware of. This is contrasted to long-term memory which is more extensive and relatively permanent. Information in working memory is lost if there is too much information to deal with, if it is not rehearsed (Peterson and Peterson 1959), or if it does not become assimilated into long-term memory (Sweller 2005). The novel information that students interact with in most classroom settings is dealt with in working memory, and therefore the limits of working memory have an impact on learning (Sweller 2005).

2. *Cognitive Schemata*: Long-term memory storage, according to CLT, is accomplished through the production of schemata. In these "cognitive constructions", multiple elements of information which would be dealt with individually in working memory are integrated into single structures relating multiple elements (Schnotz and Kürschner 2007). These schemata can interact with the novel inputs dealt with in the working memory (Sweller and Chandler 1994).

3. *Cognitive Load:* CLT posits that learners allocate cognitive resources to different tasks and "cognitive load" refers to the resource requirement placed on working memory for a particular task (Sweller and Chandler 1994).

Much early work by CLT advocates involved determining the different ways to categorize the types of cognitive load placed on working memory. The two original categories developed were intrinsic load and extraneous load. Intrinsic load refers to the "natural complexity" of a particular learning situation or activity (Schnotz and Kürschner 2007). Sweller and Chandler (1994) describe this type of load in terms of element "interactivity". Interactivity refers to the degree to which the different elements held in working memory interact. Tasks that require learners to simultaneously process many different elements of information in order to comprehend the information encountered would have high interactivity. The individual elements in these tasks could not be understood individually. However, tasks in which individual elements are more independent of each other have low interactivity. The higher the interactivity, the greater the intrinsic load (Sweller and Chandler 1994). Importantly, the expertise level of the learner is the deciding factor as to whether a learning situation involves multiple elements or not, and it is possible for two students involved in the same task to experience different levels of interactivity among elements encountered and thus, have different levels of intrinsic load (Schnotz and Kürschner 2007). Extraneous load, unlike intrinsic load, is dependent upon the instructional techniques utilized, not the inherent difficulty of the activity. Since the goal of learning, in terms of CLT is to create schemata, any cognitive activity that does not increase the opportunity for schemata construction is considered extraneous load (Sweller 2005). Extraneous load may be the result of poorly designed instruction, or could be from other classroom activity that causes learner attention, and thus cognition, to be directed elsewhere.

As CLT was being clarified in the late 1980s and early 1990s (Sweller et al. 1998) attention was directed at identifying sources of extraneous load and determining ways to minimize these. The relatively fixed nature of intrinsic load meant that if instruction was to be improved, the aspect that was open to manipulation was extraneous load (Mayer and Moreno 2003). This line of research led to identification of several "effects" associated with extraneous load including the split attention effect (Sweller et al. 1998) and the modality effect (Mayer 1997). The goal of instructional design became to decrease the extraneous load in learning situations in order to decrease the overall cognitive load. However, some researchers began to note instances in which activities that increased cognitive load actually improved student learning (Paas and van Merrienboer 1994). This led to the addition of a third category of cognitive load, germane load. Germane load is now conceived of as anything that increases load on working memory in a way that improves the opportunity to develop schemata. In that these three types of cognitive load are additive, the most desirable learning situations are those in which teaching methods decrease extraneous load while increasing germane load without resulting in an overall level of cognitive load the overwhelms the working memory of the students (Sweller 2005). Learning, as conceived in CLT, would take place when the schemata in long-term memory are changed. This "schema acquisition" ultimately means that information previously processed as several individual elements in working memory is now stored as a single element in long term memory and is available for use in working memory when needed. However, since the number of elements has been reduced, the cognitive load involved when the schemata are retrieved from long-term memory for use in working memory is less than when the individual elements were acquired as novel inputs (Schnotz and Kürschner 2007; Sweller 2005).

More recent research dealing with CLT has begun to indicate a need to reconsider the role of expertise of the learner in assessing cognitive load impact. Kalyuga et al. (1998, 2003) reported when exploring some effects associated with CLT (split attention, modality, and worked example effects), some strategies that were beneficial in terms of improving learning when used with novices were not helpful for students with greater expertise. In fact, some students with high levels of expertise were disadvantaged by the strategies. Kalyuga et al. (2003) termed this the "expertise reversal effect" in discussing the finding that instructional methods appropriate for novices may lose their effect for students with greater expertise and in some cases may become disadvantages. Schnotz and Kürschner (2007) summarize the impact of this line of research as an indication that one major factor in determining the level of cognitive load placed on a learner, and ultimately the most appropriate type of instruction for that learner, is the expertise of the learner.

Viewing CLT and the instructional developments that result from the standpoint of expertise leads to several conclusions important in the discussion of how to develop appropriate lessons which initiate appropriate cognitive load. In arguing for greater emphasis on what they call "alignment", Schnotz and Kürschner (2007) point out that tasks that are too difficult, with too great intrinsic load from too much element-interactivity overwhelm the working memory, while tasks that are too simple do not challenge the working memory to a sufficient degree. The same task may present a much greater intrinsic load on an individual with low expertise than on an individual with high expertise. In addition, increased extraneous load (even potentially from elements that for an expert could be considered intrinsic elements but are unintelligible for a novice) can sabotage learning situations. Viewed from the perspective of the students with greater expertise, tasks that have mismatched alignment due to a student having a high level of expertise relevant to task demands, may, in reality, have too low an amount of intrinsic load to facilitate schemata construction, and thus learning (Schnotz and Kürschner 2007). Analogous to Vygotski's (1963) idea of the "zone of proximal development", Schnotz and Kürschner (2007) argue that the key to setting up appropriate learning environments is to decrease extraneous load while adapting intrinsic load to the expertise of the student in a way that challenges the student at a level just beyond their current level of understanding.

Cognitive Load as Applied to Multimodal Writing Tasks in Science

Previous research has indicated that when students are asked to create multimodal writing tasks in science, there is a relationship between the degree to which the modes outside of text are embedded in or integrated with the text and student performance on end of unit tests (McDermott and Hand 2012). While these studies show that students who employ more strategies such as reference to non-text modes in the text, modal placement in text, and captions to integrate all the representations perform better on end of unit assessments (McDermott and Hand 2012; Gunel et al. 2006), they also indicate that effectively embedding modes in text is not a natural skill for students. In order to facilitate greater use of effective embeddedness strategies as a way to improve student integration and ultimately improve student understanding, the authors have developed an "Embeddedness Encouraging Lesson" (EEL) designed to help students gain awareness of issues related to improving multimodal integration in written products.

As previously mentioned, students engaged in multimodal writing tasks about science topics face the dual cognitive issues of learning about science concepts and learning about effective ways to integrate multiple modes as they communicate. CLT would predict that not only will the load requirements for each individual aspect combine in an additive fashion to determine the ultimate load on working memory, but that the expertise of each student relative to each aspect would also impact the cognitive load. Due to the fact that the embeddedness lesson employed has not been, up to this point, altered for different grade levels, the intrinsic load that characterizes it would be fixed. It would seem likely then that students at particular grade levels would have a general level of expertise that would be a "best fit" in terms of the degree to which the embeddedness lesson challenged yet did not overwhelm their working memory. In contrast, at some grade levels, students would likely be sub-challenged by the lesson or overwhelmed by the lesson and not benefit from the lesson in terms of improving degree of embeddedness in their writing. If instructional intervention is needed to help scaffold skill development in embedding multiple modes, it is important to determine the characteristics of the interventions that interact most effectively with student expertise levels. This study set out to begin this exploration by examining the effects of the embeddedness lesson on students ranging from grades 6 to 11. The research questions guiding this study included:

- Does engaging students in an EEL to promote appropriate multimodal competency impact students from different grade levels differently?
- Does a similar relationship between effective multimodal writing and conceptual understanding as demonstrated by end of unit test performance exist at different grade levels?

Methods

The research methods employed were modeled on multimodal study protocols previously utilized in classroom research settings and described by McDermott and Hand (2012). Subjects were members of pre-existing science courses in grades 6–11. Teachers who had been instructed in the use of multimodal writing tasks by the lead author were instructors for the participating students at two different sites. The use of pre-existing classes, while not ideal from a research perspective, was necessitated by the practical constraints of the school schedule. In order to account for potential pre-existing differences between groups and test for homogeneity of treatment and control groups, baseline measures of students were obtained prior to the initiation of study procedures. Characteristics of each site and of the students at each site are described further in the following section.

Participants

Site One Site one students attended a small middle school in the Midwestern United States with approximately 200 total students. One hundred and three total students in grades 6–8 were members of the classes in which research procedures were undertaken. Each grade level consisted of two separate sections of students, one of which was randomly assigned to the treatment condition and the other to the control condition. Grade 6 students (n = 15 in the treatment condition, n = 17 in the control condition) studied the topic of the nervous system during research procedures. Grade 7 students (n = 19 in the treatment condition, n = 23 in the control condition) participated in a unit focused on biomes and grade 8 students (n = 16 in the treatment condition, n = 13 in the control condition) studied a unit dealing with astronomy. Each unit was approximately 3 weeks in length and the same teacher instructed all units for all grade levels.

Site Two Students at site two attended a small high school in the Midwestern United States with approximately 220 students. Seventy-one total students participated in study procedures. Participants included grade 10 students enrolled in one of three general biology course sections and grade 11 students enrolled in one of three general chemistry course sections taught at the school. All classes were taught by the same teacher. One biology section was randomly assigned to the treatment condition (n = 11) and two were assigned to the control condition (n = 23). Study procedures took place during a unit on the human genome and a second consecutive unit on transgenic organisms. Two chemistry sections were randomly assigned to the treatment condition (n = 25) and one to the control condition (n = 12). Chemistry units during study procedures included atomic structure and electron configuration. All units were approximately 2 weeks in length.

Students at both sites and in all classes progressed through the same protocol in terms of research procedure. This protocol is described in detail in the next section.

Research Procedure

Baseline Measures Prior to the initiation of study procedures, baseline measures of student achievement were collected at each site. At site one, student final grades from the previous semester in their science course were used as a baseline measure of science understanding, while at site two, scores from the end of unit assessment for the unit preceding the study procedures were utilized. Baseline measures were used to test for homogeneity between treatment and control conditions prior to the intervention utilized in this procedure.

Embeddedness Encouraging Lesson (EEL) At the beginning of the first unit in which multimodal writing tasks were to be assigned, each treatment class received a specific lesson designed to highlight strategies used to embed multiple modes of representation within writing tasks. This lesson was termed the EEL and the outline can be found in Appendix 1. Teachers at both sites collaborated with the lead author and utilized the outline provided to attempt to ensure uniformity in the teaching of the EEL. To initiate the lesson on embeddedness, teachers provided students with an example of a unimodal (text only) communication on the scientific topic that was currently being covered in their class. Prior implementation of this EEL indicated that providing a text only description can initiate student consideration of strategies that would improve science communication, including the use of modes other than text (McDermott 2009). In addition, previous studies suggest if this unimodal example is written at a challenging reading level with a significant amount of new vocabulary, students are more likely to recognize the benefit of communication including multiple ways to represent science concepts in addition to text.

A whole class discussion was then initiated in which students analyzed the unimodal example on (1) how it effectively communicates the intended science topic,

(2) how the effectiveness might be improved if the intent of the author is to instruct the reader about the science concept and (3) how the author could use other ways to represent information (ex. pictures, diagrams, graphs, tables, etc.). The recognition of the benefits of creating multimodal rather than unimodal communication was the goal of this portion of the EEL. This was followed by student exploration of different science texts to find samples of how authors integrate text and other modes. Common strategies identified by students that authors communicating about science use include: (1) placing modes other than text near text that refers to them, (2) complete descriptions of modes other than text in the text itself, (3) captions added to modes other than text and (4) modes that are designed by the author (rather than simply copied from another source). The intent of this portion of the lesson is to help students realize the benefit of this greater embeddedness when attempting to communicate scientific concepts in a multimodal way. From the student generated list of strategies, treatment groups created a rubric designed to evaluate any piece of science communication to determine how well the different modes have been integrated. Students in the treatment groups used these class generated rubrics to assess their own writing later in the research procedure. The entire EEL lasted one full class period (approximately 55 min). Control classes did not receive the EEL and did not create the embeddedness rubric.

Unit Procedures Following the EEL with the treatment groups, unit procedures at each grade level were identical for both treatment and control sections. At an appropriate time during each unit, instructors assigned an identical writing activity to both the control and experimental groups. While the topics and type of writing involved varied across grade levels, the instructions for all writing tasks were consistent at all sites in requiring that students utilize at least one mode other than text in their product and create a magazine article that targeted 4th grade students as the audience.

Writing activities from all students (treatment and control) were collected and scored by the instructors for classroom purposes. All writing samples were then assessed by researchers with an embeddedness assessment rubric that had been utilized in previous multimodal writing studies (McDermott and Hand 2010b). This rubric will be described in greater detail in the next section. At the conclusion of each unit, all students, regardless of treatment condition, were given identical teacher created end of unit assessments designed to assess understanding of unit concepts. At site two only, all sections repeated this process for a second consecutive unit. An overview of the research design is found in Table 10.1.

Data and Analysis

Writing Sample Measures All students' writing products were analyzed using a researcher generated rubric designed to measure how well students embedded multiple modes of representation within text. This rubric was initially developed as a

Table 10.1 Overview of research design

Treatment	Control
Baseline score determined	Baseline score determined
Embeddedness encouraging lesson and construction of embeddedness checklist	
Normal unit procedures (discussion, labs)	
Writing assignment: Requirements: must include at least ONE mode of representation other than text (students assess own writing with embeddedness checklist)	Writing assignment: Requirements: must include ONE mode of representation other than text
Teacher created end of unit assessment	Teacher created end of unit assessment

part of a pilot study that was foundational for the research reported here (McDermott and Hand 2008). The rubric was developed through consultation among a group of international researchers engaged in a series of multimodal studies as a part of several current research programs, as well as through consultation with current literature related to the use of multiple modes of representation. Four sub-category scores measuring particular characteristics of student writing were developed. These sub-categories could then be combined to provide overall measures of how well students embed multiple modes with text in their writing. Scoring within each sub-category and the method of combining sub-category scores to provide overall embeddedness measures will be discussed in the following sections.

Sub-category One: Text Production Score (TPS)

The first writing assessment sub-category was text production. Four main areas were assessed here including whether the text covered the required topics from the assignment, was accurate, was complete, and was grammatically correct. A score scale of 0–3 was used with a score of 0 indicating lack of this characteristic, a score of 1 indicating this characteristic was somewhat present, a score of 2 indicating this characteristic was mostly present, and a score of 3 indicating this characteristic was completely present. A list of topics required in each unique writing assignment was generated by instructors to assess appropriate student coverage. Accuracy was assessed by noting inaccurate scientific statements and moving student scores down one level for each inaccuracy. For completeness, assessment involved determining whether each required topic was thoroughly discussed based on suggestions from the classroom instructors related to what information was necessary to cover each topic. Grammatical correctness was determined by moving student scores down one level for each grammatical inaccuracy. These four factor scores were combined and reported as the text production score (TPS).

Sub-category Two: Modal Representations Score (MRS)

The second sub-category was used to assess writing related to the use of modes other than text. This category had two main parts. First, one point was awarded for each appropriate representation other than text utilized. Secondly, one point was awarded for each different type of modal representation other than text used. For example, two pictures and three charts appropriately used resulted in two points for types of modal representations, and five points for total representations. Students were also awarded one point in this category for each of the required topics for the writing assignment that they utilized a mode outside of text to refer to. A student who developed two modes other than text referring to the central nervous system and one mode referring to the peripheral nervous system would receive two points for topics referred to. The combination of these three scores resulted in the MRS.

Sub-category Three: Average Embeddedness Score (AES)

An average embeddedness score (AES) was determined for each piece of multi-modal writing. This score was generated by assessing each instance of use of a mode outside of text in the student writing individually with a checklist containing several key factors identified in prior multimodal research as crucial for encouraging embeddedness. The key factors assessed included the accuracy of the modal representation (no scientific inaccuracies), completeness of the modal representation (did not leave out information), placement of mode next to the text that referred to it, reference to mode in the written text (used a phrase such as "see Fig. 10.1"), presence of a caption with mode, and whether the mode was an original item created by the author or one copied from another source (such as cutting and pasting on a computer). One point was awarded for each strategy employed by the author. An embeddedness score for each unique mode other than text was calculated, and then a total embeddedness score for all instances of modal use was calculated by adding all modal embeddedness scores together. The average embeddedness score was found by dividing this total embeddedness score by the number of modal representations other than text.

Cohesiveness Score (CS)

Previous data collected in research studies involving multimodal writing (particularly qualitative data) indicated that while some students employed embeddedness strategies as a means of linking different representations of the same ideas together, other students simply employed the strategies as a way to fulfill requirements of the project. These students did not actively attempt to create a well-integrated product;

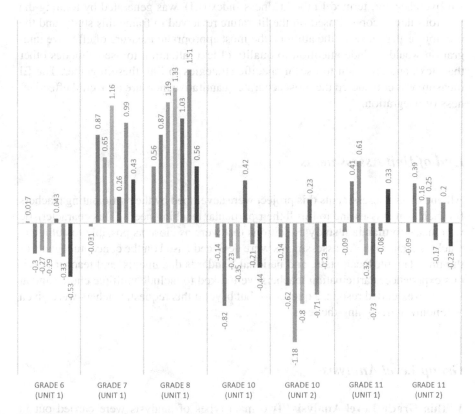

EFFECT SIZE CALCULATIONS FOR ALL LEVELS

■ BASE ■ TPS ■ MRS ■ AES ■ CS ■ EI ■ TEST

Fig. 10.1 Graphical representation of effect size data for all levels

rather they were using strategies because they had been instructed to do so. From the standpoint of the utilization of embeddedness strategies encouraging effective translation leading to conceptual development, the rote employment of strategies would not be the most likely route to effective cognition. As one way to assess use of strategies for integration purposes as opposed to use of strategies for requirement fulfillment, a fourth sub-category measure was introduced in this research project. The cohesiveness score (CS) was a measure of how well the entire product the student created was interconnected. In this category, student products were assessed in terms of whether or not the text was tied to the alternative modes throughout the entire paper, whether students carried a specific example throughout the entire product, whether individual paragraphs or sections of the total product were linked together, and whether a main conceptual idea was continually referred to. Each of these characteristics was scored as 0 (not at all), 1 (limited attempt), or 2 (consistent throughout). These scores were added to form the CS.

Embeddedness Index (EI)

An overall score, termed Embeddedness Index (EI), was generated by totaling the previous three scores. Based on the literature reviewed to frame this study, and the previous experiences of the authors, the most appropriate measure of effective integration would include attention to quality of text, attention to use of modes other than text, and attention to use of specific strategies to link these together. The EI measure was considered the most accurate quantitative measure of overall effectiveness of integration.

End of Unit Assessments

All end of unit assessments this project were developed by the participating teachers for their own classes and to fulfill their particular teaching needs. This characteristic was meant to maintain study conditions that were as close as possible to authentic teaching situations. All assessments were designed based on the conceptual goals of the units taught, local, state, and national standards documents, and teacher previous experience. Participating teachers were asked to include multiple choice and at least two extended response questions, but beyond this request, teachers were given autonomy in designing the tests.

Group Level Analysis

Within Grade Level Analysis Two main types of analysis were carried out in regard to data collected as a part of this study. Comparison of means using t-tests between treatment and control groups on writing measures and end of unit test measures for each grade level provided information related to impact of the embeddedness lesson on student writing and student understanding. Baseline scores were also analyzed using t-tests to determine whether or not differences existed in science understanding between treatment conditions at each grade level prior to the study.

Between Case Analysis Effect size calculations, as Thalheimer and Cook (2002) point out, allow for comparison across several independent studies. Because of the unique characteristics of each grade level in this study, it was determined that each level can be viewed as a separate study. Therefore, effect size calculations are helpful in determining whether similar effects from treatment were present across all grade levels. Cohen's d values were calculated for all scores described previously. Cohen (1992) provides a standard categorical scheme for classifying effect size calculations, with effect sizes below .20 categorized as negligible, effect sizes between .20 and .49 categorized as small, effect sizes of .50–.79 as medium, and effect sizes of .80 and greater as large. These categories were utilized in this study.

Analysis at the Individual Level Analysis at the level of individual students was explored through calculation of correlation values. Although correlation measures cannot indicate causality, Thompson, Diamond, McWilliam, Snyder, and Snyder (2005) point out that correlation data can be used in conjunction with other statistical analyses to provide some evidence of benefit from an intervention. "Logic based" analysis is a term used by Thompson et al. (2005) to describe a combination approach in which a variety of data analysis techniques are utilized in conjunction with a logic backed theory to evaluate a particular intervention strategy. Data describing relationships at the level of the individual student, regardless of treatment condition, is included here to support claims emerging from group level analysis data. In this study correlations measuring relationship between student writing assessment scores and performance on end of unit exams were of particular interest.

Results

As previously mentioned, the characteristics of each unique grade level were deemed different enough to warrant separate consideration of results. Of particular interest in this regard was whether or not the presence of a lesson specifically designed to encourage effective embedding of multiple modes of representation within text impacted characteristics of student writing as measured by researcher created rubrics and student conceptual understanding as measured by end of unit assessments. In addition, study procedures were designed to note whether quality of writing characteristics, regardless of treatment condition, was related to conceptual understanding. After these issues were explored at each grade level, comparisons were made across grade levels. Thus, this results section is organized in a manner that first reports results from each grade level and then discusses results across grade levels. The results presented here were collected using the measures described in the previous section. Abbreviations utilized throughout this section are provided in Table 10.2.

Table 10.2 Abbreviations utilized in results

Abbreviation	Characteristic
BASE	Baseline science competency measure
TPS	Text production score
MRS	Modal representations score
AES	Average embeddedness score
CS	Cohesiveness score
EI	Embeddedness index
TEST	End of unit evaluation score

Grade Level Analysis

Grade 6 Grade 6 data is summarized in Table 10.3. The baseline competency measure was not significantly different between treatment groups indicating no difference in science understanding prior to study procedures. Although the control group had higher mean scores for all writing measures except CS and for the end of unit test, no significant differences were detected between groups on any measure.

Individual data from Grade 6 is reported in Table 10.4. Pearson correlations were calculated between all writing measures and end of unit test performance, regardless of treatment condition. For three measures, TPS, CS, and EI, a significant, positive, correlation was detected at $p < .01$.

Grade 7 Group data for Grade 7 is reported in Table 10.5. Four significant differences all favoring the treatment groups were detected. The treatment group (M=9.78, SD=1.87) had significantly higher TPS (p=.007, t=−2.82) than the control group (M=8.13, SD=1.91). Treatment (M=8.05, SD=1.26) also had significantly higher MRS (p=.048, t=−2.043) than the control group (M=8.23, SD=1.33). The treatment group (M=11.34, SD=2.93) outperformed (p=.001, t=−3.704) the control group (M=7.36, SD=3.83) on AES and the treatment group (M=36.44, SD, 5.83) had significantly higher EI scores (p=.003, t=−3.136) than the control group (M=28.76, SD=9.26).

Individual Grade 7 data is summarized in Table 10.6. No significant correlations were detected between any writing characteristics and end of unit test scores.

Table 10.3 Summary of grade 6 group data

Unit	Question type/writing measure	Experimental (n=15)		Control (n=17)	
		M	SD	M	SD
	BASE	89.40	10.09	89.23	8.71
1	TPS	8.33	2.96	9.05	1.85
	MRS	5.46	1.12	5.88	1.83
	AES	7.90	1.33	8.41	2.09
	CS	7.33	1.34	7.26	1.85
	EI	29.04	3.25	30.61	5.94
	TEST	28.36	4.95	30.70	3.70

Table 10.4 Grade 6 writing characteristic correlations with test score (n=32)

Test score correlation with	Pearson correlation
TPS	.454**
MRS	.271
AES	.127
CS	.519**
EI	.539**

*p < .01

Table 10.5 Summary of grade 7 group data

Unit	Question type/writing measure	Experimental (n=19)		Control (n=23)	
		M	SD	M	SD
	BASE	91.15	4.48	91.3	4.28
1	TPS	9.78*	1.87	8.13	1.91
	MRS	8.05*	1.26	6.56	2.95
	AES	11.34*	2.93	7.36	3.83
	CS	7.26	1.93	6.69	2.95
	EI	36.44*	5.83	28.76	9.26
	TEST	47.10	8.52	43.50	8.39

*p<.05

Table 10.6 Grade 7 writing characteristic correlations with test score (n=42)

Test score correlation with	Pearson correlation
TPS	.206
MRS	.074
AES	.289
CS	−.073
EI	.182

Table 10.7 Summary of grade 8 group data

Unit	Question type/writing measure	Experimental (n=16)		Control (n=13)	
		M	SD	M	SD
	BASE	93.81	6.54	90.30	6.04
1	TPS	9.62*	1.82	7.92	2.10
	MRS	8.12*	1.20	6.61	1.32
	AES	10.50*	2.20	7.92	1.61
	CS	8.50*	1.03	7.07	1.65
	EI	36.75*	3.93	29.53	5.50
	TEST	46.56	9.65	40.84	10.86

*p<.05

Grade 8 Grade 8 student data is summarized in Table 10.7. For all writing measures, the treatment group had significantly higher scores than the control group. The treatment group (M=9.62, SD=1.82) had significantly higher TPS (p=.027, t=−2.337) than the control group (M=7.92, SD=2.10). The treatment group (M=8.12, SD=1.20) had significantly higher MRS (p=.003, t=−3.210) than the control group (M=6.61, SD=1.32). Treatment (M=10.50, SD=2.20) had significantly higher AES (p=.002, t=−3.509) than control (M=7.92, SD=1.61). Also, the treatment group (M=36.75, SD=3.93) had higher EI scores (p=.000, t=−4.112) than the control group (M=29.53, SD=5.50).

Table 10.8 Grade 8 writing characteristic correlations with test score (n = 29)

Test score correlation with	Pearson correlation
TPS	.518**
MRS	.318
AES	.384*
CS	.534**
EI	.552**

* = significant at p < .05 level

** = significant at p < .01 level

Table 10.9 Summary of grade 10 group data

Unit	Question type/writing measure	Experimental (n = 11)		Control (n = 23)	
		M	SD	M	SD
	BASE	28.09	9.87	29.37	7.80
1	TPS	10.18	1.33	11.13*	0.97
	MRS	7.72	1.10	8.00	1.35
	AES	7.11	1.91	7.79	1.81
	CS	10.09	1.45	9.26	2.36
	EI	35.11	5.09	36.16	4.93
	TEST	31.00	4.02	32.63	3.28
2	TPS	9.91	1.58	10.78	1.20
	MRS	7.00	1.18	8.30*	1.02
	AES	6.95	1.86	8.39*	1.74
	CS	9.81	1.33	9.43	1.97
	EI	33.68	4.56	36.92	4.51
	TEST	51.72	7.64	53.70	9.40

*p < .05

Individual data from Grade 8 is summarized in Table 10.8. One writing characteristic (AES) was significantly positively correlated with test scores at the p < .05 level, while three others, TPS, CS, and EI, were significantly positively correlated at the p < .01 level.

Grade 10 Students in grade 10 classes participated in two consecutive units in which multimodal writing tasks were assigned. For the first unit, only one category contained significantly different scores between treatment conditions. For TPS, the control group (M = 11.13, SD = .97) outperformed (p = .024, t = 2.367) the treatment group (M = 10.18, SD = 1.33). For unit two, the control group (M = 8.30, SD = 1.02) outperformed the treatment group (M = 7.00, SD = 1.18) on MRS (p = .002, t = 3.315). In addition, the control group (M = 8.39, SD = 1.97) significantly outperformed the treatment group (M = 6.95, SD = 1.86) on AES (p = .035, t = 2.236). Grade 10 group data is summarized in Table 10.9.

Table 10.10 Grade 10 writing characteristic correlations with test score (n = 34)

Test score correlation with	Pearson correlation
TPS (Unit 1)	.263
MRS (Unit 1)	.429*
AES (Unit 1)	.461**
CS (Unit 1)	.466**
EI (Unit 1)	.545**
TPS (Unit 2)	.530**
MRS (Unit 2)	.470**
AES (Unit 2)	.060
CS (Unit 2)	.592**
EI (Unit 2)	.525**

* = significant at p < .05 level
** = significant at p < .01 level

Table 10.11 Summary of grade 11 group data

Unit	Question type/writing measure	Experimental (n = 25)		Control (n = 12)	
		M	SD	M	SD
	BASE	28.24	11.84	29.41	13.28
1	TPS	10.80	1.19	10.25	1.48
	MRS	9.32	0.85	8.67	1.23
	AES	7.40	1.43	7.91	1.66
	CS	9.52	1.66	10.50	0.90
	EI	37.04	3.41	37.33	3.56
	TEST	54.58	7.28	51.97	8.37
2	TPS	10.52	1.56	10.00	1.04
	MRS	9.00	1.22	8.83	0.93
	AES	8.64	1.91	8.19	1.59
	CS	9.28	1.74	9.58	1.62
	EI	37.44	4.70	36.60	3.58
	TEST	53.90	12.39	56.37	8.29

*p < .05

Individual data for grade 10 is summarized in Table 10.10. Two scores, TPS in unit one and AES in unit two, were not significantly correlated with tests scores for the respective units. One score, MRS for unit one was significantly correlated to test score at the p = .05 level. The rest of the scores for each unit were significantly correlated with test scores at the p = .01 level.

Grade 11 For grade 11, no significant differences were detected between treatment and control groups on any measures. Group 11 group data is summarized in Table 10.11.

Table 10.12 Grade 11
writing characteristic
correlations with test score
($n = 37$)

Test score correlation with	Pearson correlation
TPS (Unit 1)	−.149
MRS (Unit 1)	−.017
AES (Unit 1)	.242
CS (Unit 1)	−.117
EI (Unit 1)	−.005
TPS (Unit 2)	.335*
MRS (Unit 2)	.310
AES (Unit 2)	.163
CS (Unit 2)	.138
EI (Unit 2)	.312

* = significant at $p < .05$ level

For individual data with grade 11 students, only one measure, TPS for unit 2, was significantly positively correlated with test scores. No other significant correlations were detected. Table 10.12 summarizes grade 11 individual data.

Cross Case Analysis

The results described at the level of individual grades can be compared across grades by utilizing effect size data. Table 10.13 summarizes the effect size data from all grade levels and Fig. 10.1 represents this data graphically. For each measure, a Cohen's d value is reported, along with an effect size classification based on the scheme discussed in the methods section. Positive effect sizes indicated greater performance by the treatment group.

Discussion

Expectations Emerging from Cognitive Load Theory

As discussed in the literature review, previous research in limited science classrooms has indicated benefit when students effectively integrate multiple modes of representation in writing tasks. However, the fact that this type of integration was not naturally employed by students led to the development of an EEL aimed at improving students' ability to integrate multiple modes. Along with the development of this lesson arose awareness that this additional information may create additional cognitive load for students involved in these tasks. Current research related to CLT would seem to indicate that this additional load could be either beneficial or disadvantageous for students based at least partially on the expertise of the

Table 10.13 Effect size calculations for all measures by grade level

Grade	Unit	Measure	Cohen's d	Effect size
6		BASE	0.017	Negligible
		TPS	−0.3	Small
		MRS	−0.27	Small
		AES	−0.29	Small
		CS	0.043	Negligible
		EI	−0.33	Small
		TEST	−0.53	Medium
7		BASE	−0.031	Negligible
		TPS	0.87	Large
		MRS	0.65	Medium
		AES	1.16	Large
		CS	0.26	Small
		EI	0.99	Large
		TEST	0.43	Medium
8		BASE	0.56	Medium
		TPS	0.87	Large
		MRS	1.19	Large
		AES	1.33	Large
		CS	1.03	Large
		EI	1.51	Large
		TEST	0.56	Medium
10		BASE	−0.14	Negligible
	(Unit 1)	TPS	−0.82	Large
		MRS	−0.23	Small
		AES	−0.35	Small
		CS	0.42	Small
		EI	−0.21	Small
		TEST	−0.44	Small
	(Unit 2)	TPS	−0.62	Medium
		MRS	−1.18	Large
		AES	−0.8	Large
		CS	0.23	Small
		EI	−0.71	Medium
		TEST	−0.23	Small
11		BASE	−0.09	Negligible
	(Unit 1)	TPS	0.41	Small
		MRS	0.61	Medium
		AES	−0.32	Small
		CS	−0.73	Medium
		EI	−0.08	Negligible
		TEST	0.33	Small
	(Unit 2)	TPS	0.39	Small

(continued)

Table 10.13 (continued)

Grade	Unit	Measure	Cohen's d	Effect size
		MRS	0.16	Small
		AES	0.25	Small
		CS	−0.17	Small
		EI	0.2	Small
		TEST	−0.23	Small

students. Combining the current research on CLT with the current pedagogical progression that is used in relation to the multimodal writing tasks being explored here leads to two main predications. First, if the embeddedness encouraging lesson provides an appropriate match between student expertise and intrinsic load, student understanding relative to the skill of appropriately embedding modes in text should be enhanced. One way this enhancement would be exhibited would be through higher scores on the researcher developed measures of writing characteristics discussed previously. In practical terms for this study, this would likely be manifested by significantly higher scores in terms of writing scores for treatment classes (who received the EEL) as compared to control classes (who did not). Secondly, if student creation of multimodal writing products itself is a task which is appropriately matched with student expertise, then performance on end of unit tests should be related to the measures of writing characteristics. Significant correlations between writing scores and test performance would be one indication of this relationship in this particular study.

Appropriateness of EEL

A definite trend develops when analyzing the differences between control and treatment groups in relation to the writing characteristics measured. In grade 6, no significant differences appear between treatment and control groups. In grade 7, however, significant differences favoring the treatment group are indicated for scores related to text, use of modes and the overall measure of Embeddedness Index, while for grade 8 students, all writing characteristics were significantly higher for the treatment students who had received the EEL. In the older grades, the advantage for the treatment students disappears and in the case of the grade 10 students, the intervention of the EEL may have even disadvantaged the students as the control group outperformed the treatment group on text performance, use of modes, and average embeddedness score. This pattern would suggest that the current format and rigor of the EEL has an intrinsic load that is most appropriately matched with the expertise of grade 7 and grade 8 students. For grade 6 students, the lesson may have had too great a load component and overwhelmed the working memory of students.

For grade 10 and 11 students, the expertise reversal effect may have been present. It is possible that the intrinsic load of the lesson is not challenging enough for these students to utilize the material as a way to improve understanding. In addition, Schnotz and Kürschner (2007) suggest that one way extraneous load may be increased is through wasted (or perceived wasted) effort and time. If the grade 10 and 11 students perceived the lesson on embeddedness as unnecessary because they felt they were already aware of these skills, this may have increased extraneous load and not contributed to learning.

Appropriateness of Multimodal Tasks

A second factor related to the use of multimodal writing tasks relates to the general potential for these tasks to improve conceptual understanding. While improving the ability to create effective multimodal products and employ integration strategies is important, the ultimate goal of science education is to improve student conceptual understanding. It is conceivable that the EEL could be appropriately matched to expertise in a way that promotes better writing skill, but that the use of the actual assignment in which students create multimodal products does not create an appropriately matched cognitive load situation for the encouragement of developing understanding about targeted science concepts. The data collected here indicates some variability in the relationship between writing scores in general (regardless of treatment condition) and test performance. Grades 8 and 10 exhibited the greatest number of writing characteristics that were significantly correlated with test performance, indicating that for these grade levels, the task of creating well-integrated products is appropriately challenging and may benefit conceptual benefit. For grade 11 students, significant relationships virtually disappeared, which may be another indication of a sub-challenged level of intrinsic load in general from these types of assignments or a perceived lack of importance on the part of the students. Grades 6 and 7 showed some significant correlations, but not as consistent as in grades 8 and 10. This may indicate that the task overwhelmed some of these students in a way that too much cognitive attention was needed for attending to rhetorical issues which did not allow for attention to the conceptual issues.

Limitations and Further Questions

Several limitations exist in relation to this study and some of these suggest new avenues of research. As with any educational study set in authentic school settings, the degree to which teacher implementation of the research protocol is standardized is always open to debate. It is quite likely that the different teachers involved did not teach in an identical manner, regardless of the attempts to standardize lesson delivery. Differences in the implementation of the lessons may have impacted results. Another limitation relates to the standardization of the topic. The topics dealt with at the different grade levels were not standardized. It is quite likely that different topics within the curriculum for a particular grade level have different degrees of intrinsic load. It is possible that if the topics of study utilized in this research project had vastly different intrinsic loads, these differences may have impacted outcomes more than the absence or presence of the EEL. It is also possible that particular types of modes provide better "fits" for particular topics. In addition, some research has indicated that the sequence of required modal use may impact student learning. Therefore, further research may explore these issues in relation to cognitive load. Would the required use of a particular type of mode be more beneficial in terms of creating an optimal cognitive load situation than the use of other types of modes or even the allowance of free choice by students for modes?

A third limitation relates to previous findings indicating that consistent opportunity to engage in modal use may increase not only effective integration of all modes but also conceptual benefit from the production of multimodal products (McDermott 2009). In this study, only grade 10 and 11 students were given multiple opportunities to create multimodal products. It is possible that for younger aged students, while the first experience with multimodal writing and the EEL overwhelms the working memory for the initial unit used, if students were given a second opportunity, the previous experience might improve expertise and therefore promote more beneficial integration in the second experience. In addition, perhaps with the grade 8 students, some of the initial benefit may fade in progressive interactions with multimodal tasks. Another way to provide consistent opportunity to engage with multimodal communication is to refer to the EEL consistently throughout the unit as multiple representations are encountered. The research procedures for this study did not include analysis of the degree to which students in treatment classes referred to the checklist they created or multiple modes in general throughout the unit of study and if this differed from control classes. This issue of consistent referral would be a variable of interest in further study.

Finally, several further areas of research are suggested in this study. One emerging issue related to most instructional design is the issue of technology. Utilization of technological tools in the delivery of the EEL or in the production of the multimodal tasks would likely impact load issues as well as overall learning. Much of the CLT discussion has emerged from exploring how students learn from multimodal learning environments (such as when they use computer simulations). Perhaps a new line of inquiry would involve the relationship between CLT and the production of technologically enhanced multimodal products. Student motivation and perception of the necessity of the skill development involved may also play a role in cognitive load and in the benefits associated with this lesson. Student perception that the embeddedness strategies are either not important or critically important may impact the degree to which the engagement with the strategies places load constraints on the learner. It is even possible that the sequencing of the unit plans communicates to students the importance (or lack thereof) of the multimodal strategies, and therefore impacts load on working memory. In all of these cases, it is clear that further development of the most appropriate ways to design multimodal writing tasks and the embeddedness encouraging lessons associated with them is necessary. Considering CLT along with this further development can help increase opportunity for benefit emerging from the activities.

General Conclusion and Further Study

Undoubtedly, no single factor is responsible for the cognitive load placed on students from the EEL and the multimodal writing tasks. While this study begins to indicate differentiation in terms of what grade levels the lesson is most appropriate for and what grade levels the activity itself is most appropriate for, much further clarification is needed. It is more likely that several factors, including the science topic under study, the type of modes chosen by students (or required by teachers), the form of writing product called for, as well as the format of the embeddedness lesson all interact in different ways to impact cognitive load, and ultimately, learning. In addition, the number and type of follow-up opportunities students engage in during the unit of study related to multimodal communication is another variable likely to impact student performance that could be examined. The next step in making these tasks and the associated lessons a more pedagogically sound tool is exploration of these factors individually, as well as how they interact in a variety of settings.

Appendix 1

Embeddedness Encouraging Lesson (EEL) Outline

Overview A full understanding of any science concept would likely require a "multimodal" understanding of the concept in that the student would be able to represent their understanding using an assortment of modes *and* the student would be able to effectively link these multiple representations together to create a well-integrated and cohesive piece of communication. The ultimate intent of this lesson is to help students recognize that when utilizing modes other than text along with text to communicate about a scientific concept, there are strategies and techniques that can be employed to improve the overall cohesiveness of the written communication. In this lesson, students will be given the opportunity to explore common sources of communication about science to observe not only how different modes are utilized to describe science ideas but also to identify strategies that are utilized to link the different modes (including text) together. The lesson will culminate in the production of an assessment matrix that can be utilized to evaluate how well any piece of written communication has integrated multiple modes of representation.

Outline The lesson should be implemented in the classroom in a manner that fits with the overall progression of the course. In that sense, the lesson should be tailored to meet the specific needs of the specific classroom it is employed in. However, analysis of previous attempts to utilize lessons of this type have indicated that greater effectiveness in impacting student writing and ultimately student understanding are achieved when all of the following components are present:

1. **Student assessment of unimodal (text only) communication:**
   Students can be given a text only description of the topic or concept that will be discussed in the next unit of study. Ask students to analyze the description in terms of what aspects of the communication are effective and how could the effectiveness be improved if the intent of the author is to instruct the reader about the new science concept. In the course of this discussion, ideally, students will recognize that use of other ways to represent information (pictures, diagrams, graphs, tables, etc.) can be helpful. It is often beneficial to have a text only description that is fairly difficult in terms of reading level and heavy in new vocabulary so that most students will be unable to completely grasp the understanding from the description alone.

2. **Student identification of multimodal usage and strategies to link modes:**
   Once students have indicated that science communication is enhanced through the use of modes other than text along with text, students should be given the opportunity to use common sources of science information to find examples of multimodal use and examples of strategies to link modes together. Textbooks, science magazines, websites, or newspaper articles can be provided for this component of the lesson. Ask students to list all modes other than text used in the sources they observe. Discussion about these modes is an appropriate time to discuss naming of the modes. For example, students may disagree on what makes a diagram different than a picture (often students decide that a diagram has labels while a picture does not). While it is not necessary to come to agreement on what constitutes each mode, it is usually easier in future discussions if some sort of consistent naming system is used.

   The other main factor of this lesson component is for students to identify strategies that authors use to link different modes together. Common strategies that are often identified include placing modes other than text near text that refers to them, complete textual descriptions of modes in the text, captions added to modes other than text, and the modes that are designed by the author (rather than simply copied from another source). The goal is for students to develop as thorough a list as possible of common rhetorical strategies utilized by authors to tie all modes together. Again, discussion about the usefulness of different strategies is encouraged.

3. **Student creation of assessment matrix:**
   The culminating activity for this lesson is the production of a student generated matrix for assessing any piece of written communication for how well different modes are integrated and the level of cohesiveness of the entire product. The particular format of the matrix is not critical; rather, the core issue is that students

utilize the ideas about modal use and strategies to link modes that they developed in part 2 to create a practical assessment tool. An example of a student checklist created as a part of earlier lessons is included in this packet. Students will utilize this matrix to self-analyze their own writing and potentially, depending on how further lessons are structured, to analyze written communication by other authors.

4. **Student opportunities to practice using assessment matrix:**
It is often helpful if students are given the opportunity to utilize the newly created assessment tools from part 3 to authentically examine science communication. Students can review the sources they observed in part 2 or other sources. Websites describing science content may be projected for the entire class to see and each student can assess using the matrix. Discussion following this component is often focused on both the practical use of the matrix and the relative necessity or benefit of different strategies students have identified as important. In some cases, at this point in the lesson, students have been asked to create multimodal products and after exchanging with a peer, assess the products. An effective strategy if consecutive units using multimodal writing tasks are used is to have students analyze peers' work from the first unit using the matrix as a review during the second unit.

Obviously, the more this lesson and the products of this lesson are referred to throughout the overall unit, the more the ideas become a critical factor for students to consider. Greater emphasis on using different modes and connecting the different modes will likely lead to students paying more attention to these characteristics in sources they use and communication they produce. The more the encouragement of this type of communication becomes a "normal" aspect of the classroom environment, the more students will benefit from the practice.

References

Alvermann, D. (2004). Multiliteracies and self-questioning in the service of science learning. In W. Saul (Ed.), *Crossing borders in literacy and science instruction: Perspectives on theory and practice*. Newark: International Reading Association.

Bereiter, C., & Scardamalia, M. (1987). *The psychology of written composition*. Hllsdale: Lawrence Erlbaum Associates.

Cohen, J. (1992). A power primer. *Psychological Bulletin, 112*(1), 155–159.

Dorfner, G. (1999). The connectionist route to embodiment and dynamicism. In A. Reigler, M. Peschl, & A. von Stein (Eds.), *Understanding representations in cognitive sciences*. New York: Kluwer Academic/Plenum.

Emig, J. (1977). Writing as a mode of learning. *College Composition and Communication, 28*, 122–128.

Galbraith, D. (1999). Writing as a knowledge-constituting process. In D. Galbraith & M. Torrance (Eds.), *Knowing what to write: Conceptual processes in text* (pp. 139–159). Amsterdam: Amsterdam University Press.

Gunel, M., Hand, B., & Gunduz, S. (2006). Comparing student understanding of quantum physics when embedding multimodal representations into two different writing formats: Presentation format vs. summary report format. *Science Education, 90*(6), 1092–1112.

Gunel, M., Atila, M. E., & Buyukkasap, E. (2009). The impact of using multi modal representations within writing to learn activities on learning electricity unit at 6th grade. *Elementary Education Online, 8*(1), 183–199.

Hand, B. (2007). *Science inquiry, argument and language: A case for the science writing heuristic.* Rotterdam: Sense Publishing.

Kalyuga, S., Chandler, P., & Sweller, J. (1998). Levels of expertise and instructional design. *Human Factors, 40*(1), 1–17.

Kalyuga, S., Ayers, P., Chandler, P., & Sweller, J. (2003). The expertise reversal effect. *Educational Psychologist, 38*, 23–31.

Klein, U. (2001). Introduction. In U. Klein (Ed.), *Tools and modes of representation in the laboratory sciences.* Boston: Kluwer Academic.

Klein, P. (2006). The challenges of scientific literacy: From the viewpoint of second-generation cognitive science. *International Journal of Science Education, 28*, 143–178.

Kress, G., Charalampos, T., & Ogborn, J. (2006). *Multimodal teaching and learning: The rhetorics of the science classroom.* London: Continuum International Publishing Group.

Mayer, R. (1997). Multimedia learning: Are we asking the right questions? *Educational Psychologist, 32*(1), 1–19.

McDermott, M. A. (2009). *The impact of embedding multiple modes of representation on student construction of chemistry knowledge.* Ph.D. thesis, University of Iowa. http://ir.uiowa.edu/etd/253

McDermott, M., & Hand, B. (2008, July 2–5). *The impact of embedding multiple modes of representing information in writing to learn activities on student construction of chemistry knowledge.* Paper presented at the annual meeting of the Australasian Science Education Research Association conference, Brisbane.

McDermott, M., & Hand, B. (2010a, March 20–24). *Exploring the impact of embedding multiple modes of representing science information in varied classroom settings.* Paper accepted for presentation as part of a paper set on writing-to-learn activities at annual conference of National Association of Research in Science Teaching (NARST), Philadelphia.

McDermott, M., & Hand, B. (2010b). A secondary reanalysis of student perceptions while participating in non-traditional writing in science. *Journal of Research in Science Teaching, 47*(5), 518–539.

McDermott, M., & Hand, B. (2012). The impact of embedding multiple modes of representation within writing tasks on high school students' chemistry understanding. *Instructional Science, 41*, 217–246.

National Research Council. (1996). *National science education standards.* Washington, DC: National Academy Press.

Paas, F., & van Merrienboer, J. J. G. (1994). Variability of worked examples and transfer of geometrical problem solving skills: A cognitive load approach. *Journal of Educational Psychology, 86*, 122–133.

Peterson, L., & Peterson, M. (1959). Short-term retention of individual verbal items. *Journal of Experimental Psychology, 58*, 193–198.

Prain, V. (2006). Learning from writing in secondary science: Some theoretical and practical implications. *International Journal of Science Education, 28*(2), 179–201.

Prain, V., & Hand, B. (1996). Writing and learning in secondary science: Rethinking practices. *Teacher and Teacher Education, 12*, 609–626.

Schnotz, W., & Kürschner, C. (2007). A reconsideration of cognitive load theory. *Educational Psychology Review, 19*, 469–508.

Seufert, T. (2003). Supporting coherence formation in learning from multiple representations. *Learning and Instruction, 13*(2), 227–237.

Sweller, J. (2005). Implications of cognitive load theory for multimedia learning. In R. E. Mayer (Ed.), *The Cambridge handbook of multimedia learning* (pp. 19–30). New York: Cambridge University Press.

Sweller, J., & Chandler, P. (1994). Why some material is difficult to learn. *Cognition and Instruction, 12*(3), 185–233.

Sweller, J., van Merrienboer, J., & Paas, F. (1998). Cognitive architecture and instructional design. *Educational Psychology Review, 10*(3), 251–296.

Thalheimer, W., & Cook, S. (2002). How to calculate effect sizes from published research simplified methodology. Retrieved from http://www.learningaudit.com/white_papers/effect_sizes/Effect_Sizes_pdf4.pdf

Thompson, B., Diamond, K. E., McWilliam, R., Snyder, P., & Snyder, S. W. (2005). Evaluating the quality of evidence from correlational research for evidence-based practice. *Exceptional Children, 71*, 181–194.

Vygotski, L. S. (1963). Learning and mental development at school age. In B. Simon & J. Simon (Eds.), *Educational psychology in the U.S.S.R* (pp. 21–34). London: Routledge and Keegan Paul.

Waldrip, B., Prain, V., & Carolan, J. (2010). Using multi-modal representations to improve learning in junior secondary school science. *Research in Science Education, 40*, 60–85.

Yore, L. (2012). Science literacy for all – More than a slogan, logo, or rally flag! In K. C. D. Tan, M. Kim, & S. Hwang (Eds.), *Issues and challenges in science education research: Moving forward*. Dordrecht: Springer.

Chapter 11
Using a Framing Analysis to Elucidate Learning from a Pedagogy of Student-Constructed Representations in Science

Jim Carolan

Literature Review

The concept of framing has been extensively developed and utilised to inform understanding in varied disciplines including sociology, linguistics, anthropology and more recently education (Berland and Hammer 2012). Most simply stated, a frame entails the sense made by an individual of 'what is it that is going on here?' (Goffman 1974, p. 8). That is, what perspective on, or aspects of, any particular situation might the individual be sensitive to. Framing is seen to be ubiquitous with each and every participant in a situation necessarily framing it in some way (Greeno 2009). While such framings often remain personal, tacit and even subconscious (Elby and Hammer 2010) they are considered to shape the individual's response(s) to the activity and ultimately the development of their understanding (Berland and Hammer 2012; Hofer 2001). From this perspective quality learning of scientific knowledge and its production practices of representational discourse requires teaching that supports student development of consistent, coherent, productive framings in relation to these (Shemwell and Furtak 2010; van de Sande and Greeno 2012). With framing seen as a physically- and socially- dynamic, ongoing process where individuals constantly frame and re-frame their understanding (Berland and Hammer 2012) their actions, utterances and artifacts produced may be interpreted as snapshots indicative of the framings held at that moment. A series of such evidence thus provides an indication of the potential development of framings in response to sequences of activities in which the individual has participated. Such considerations are seen as having potential to add significant depth to teacher evaluations of learning (Bing and Redish 2012) and the effectiveness of their practice (Roberts 1996).

J. Carolan (✉)
La Trobe University, Melbourne 3086, VIC, Australia
e-mail: j.carolan@latrobe.edu.au

© Springer International Publishing Switzerland 2016   213
B. Hand et al. (eds.), *Using Multimodal Representations to Support Learning in the Science Classroom*, DOI 10.1007/978-3-319-16450-2_11

At any particular point the framings of an individual are influenced by the 'meta-communicative messages' they receive via the personal, social and physical resources available (Bateson 1972, p. 188). For example, as people indicate, both explicitly and more subtly, their own framing(s) of what is taking place others may respond to these signals by adjusting their own framing(s) of the situation (Berland and Hammer 2012). Different personal framings may act concurrently and include diverse but related interpretations such as the perceived objective or scope of a task and participant roles therein, through to what counts as knowledge and how this might be generated or communicated (Elby and Hammer 2010; van de Sande and Greeno 2012). For the purposes of this study three broad aspects of framing identified in the literature were considered – *positional* and *conceptual* framing developed by van de Sande and Greeno (2010), and *epistemological* framing characterized by Elby and Hammer (Elby and Hammer 2010; Hammer et al. 2005). *Positional* framing refers to how participants understand themselves and others to be related to each another, group and/or resources in an interaction (van de Sande and Greeno 2012). *Conceptual* framing as used by van de Sande and Greeno (2012) refers to ways in which participants perceive the organisation of information in a situation. *Epistemological* framing relates to the sense of the participants answer to Goffman's question "what is it that is going on here?" with respect to knowledge (Redish 2004). This includes understanding of what kinds of knowledge are useful in the activity they are engaged in terms of how answers should be constructed and validated (Berland and Hammer 2012).

While the framing aspects outlined here are not viewed as separately variable (van de Sande and Greeno 2012) these distinctions are considered valuable for identification and analysis of the frames students engage during moments of interaction as they become attuned to, coordinate and mobilize different resources for learning. For example, advances in conceptual understanding thought of as conceptual replacement from first-generation perspectives are conjectured to implicate the development of new resources for framing in a contemporary framing perspective (van de Sande and Greeno 2010; Vosniadou et al. 2008). Alignment of positional framings can be seen as enabling engagement in a community of practice – whose members have aligned epistemological and conceptual framings. Learning in this context can be seen as a further alignment of framings – both epistemological with respect to the nature of science as a knowledge producing community of practice, and conceptual with respect to how science knowledge is constituted in that practice.

Positional Framing

Positional framing refers to how participants understand themselves and others to be related to each another, group and/or resources in an interaction (van de Sande and Greeno 2012). This affects the resulting pattern of activity with respect to a subject matter and the available representational resources by defining understandings of who within the group has the right or expectation to raise and attempt to resolve questions, in what nature of representation and how (including challenge to others' ideas) and determine if, when and how they are resolved. It is from this positional framing that the student projects as an actor in the situation (Greeno and MacWhinney 2006). Positional framings are also seen here as applying to solo activity in how the individuals "position" themselves with respect their interactions with tasks or resources e.g. do they see their role to be a conveyor or a interpreter of the information?

Epistemological Framing

The term epistemological frame was coined by Redish (Redish 2004) to connect the study of personal epistemologies to the notion of framing in the sense of the participants answer to the question "what is going on here?" with respect to knowledge. This includes understanding of what kinds of knowledge are expected and/or useful in the activity (e.g. what kind of representation enables thinking, what would count as a solution to the problem for that situation, or how new knowledge is built (Hammer et al. 2005)). Epistemological framing is seen as playing a crucial role in understanding and reasoning (Hammer et al. 2005; Scherr and Hammer 2009). Elby and Hammer (2010) contend that students often manifest different epistemologies in different contexts i.e. they apply different epistemological frames.

There are noted practical connections between the positional and epistemological framings of students that have significant implications with regard to learning activities for developing science literacy. Berland and Hammer's (2012) analysis suggests that students' and teacher's social expectations (positional framings) interact significantly with their epistemological expectations (framings). They posit that they use the former 'as tacit proxies for the latter' (Berland and Hammer 2012, p. 89). For example, Van de Sande and Greeno (2012) noted that unless they 'were positioned in relation to the subject matter as being entitled to conduct inquiry' participants' tacit assumptions that the 'knowledge needed to complete the tasks could be gained through inquiry'(van de Sande and Greeno 2012, p. 41) could not be enacted. The development of student epistemologies regarding the nature of science thus appears to be strongly influenced by the framing of roles in classroom learning activities in a case of social mediation paralleling that of the scientific enterprise (Shemwell and Furtak 2010). This has implications for development of a contemporary epistemological framing of science knowledge production practices by students. If an ideal learning activity has students engage as a "community of peers" amongst whom ideas are represented as arguments to build warrants and judge their merit (Ford and Forman 2006; Shemwell and Furtak 2010) the success of epistemological learning from such an activity could be seen as dependent on the style and level of participation. For example, it has been observed that a learner may take a *depictive* perspective in that they frame themselves as an observer of the construction of the representation/explanation or an *enactive* perspective whereby they frame themselves as a participant in the constructive activity (MacWhinney 2005). Along similar lines, Berland and Hammer (2012) noted that student framings where the topic of discussion is expected to be identified by the teacher aligns with idea-sharing (depictive) interactions, whereas student framing with student topic identification aligns with argumentative (enactive) interactions. Thus such a positional framing of the student regarding their conceptual agency in the construction of scientific representations in the classroom can affect their development of productive epistemological framings.

Conceptual Framing

Conceptual framing as used by van de Sande and Greeno (2012) refers to ways in which participants perceive the organization (representation) of information in a situation. In this view of conceptual framing (as for positional framing) an alignment of conceptual frames is necessary for achievement of "mutual understanding" regarding that situation. In the classroom this generally correlates to the student conceptual understanding (schema or frame) being coherent with the scientifically accepted understanding (schema or frame) expressed by the teacher or other source (e.g. text or web). For learning to have generality, however, the learner must become 'attuned to constraints and affordances that are invariant or similar across transformations of the learning situation' (Greeno 2009, p. 271). That is, the student must develop, recall and apply a conceptual frame that consistently identifies continuities of features across different situations (i.e. structural coherence). Features are conceived here as the organization or "organizability" of information whose recognition in turn entails a conceptual structure or schema. In learning students are cued (or guided) to recognize the productive nature of such a conceptual framing (applied) over multiple examples. Generality is achieved by the students as they attend to 'aspects of specific situations in relation to structures that are general' (Greeno 2009, p. 274).

Van de Sande and Greeno's (2012) use of conceptual framing also reflects the representational nature of information in science – what is selected for representation, for what purpose and how. Consider Mendeleev's representation of the periodicity of the properties of the known elements. An integral aspect if his construction of his periodic table was his choice to focus only on the measurable elemental properties he perceived as the most relevant for the purpose – firstly atomic weight, then valency and isomorphism (Weisberg 2007). Van de Sande and Greeno refer to conceptual framings as 'relations between different aspects of information in the situation, including those that are attended to and those that are not attended to, and those that are in the foreground and those that are in the background of attention' (van de Sande and Greeno 2012, pp. 3, 4). This reflects representational distortions in the scientific process of construction and use of models as explanatory tools (Woods and Rosales 2010). This over- and/or under-representation of aspects of empirical phenomena are purposeful idealizations as means of conjecture (Woods and Rosales

2010). Greeno and Van de Sande tie conceptual growth of the individual to the increase in their capability to construct framings that 'use the constraints and affordances that constitute the meanings of concepts' (Greeno and van de Sande 2007, p. 16). Scientific concepts are instantiated by the representational relations used to make meaning in practice and so developing productive framings entails developing competence in such representational practices. Recognition (tacitly through representational practice or overtly) of science knowledge claims as fundamentally a representational practice of argument thus ties conceptual and epistemological framing in higher levels of representational competence. This involves students positioned as explainers in the classroom, utilizing representational affordances and constraints to explicitly make claims.

In practice the effective development of scientific literacy can be seen to require an alignment of positional (i.e. the student accepts the role as constructor of scientific argument), conceptual (i.e. the student identifies the appropriate representational affordances and constraints) and epistemological framings (i.e. the student uses these to explicitly construct the argument).

Teacher Framing

A focus on representation construction by students has potential to clarify teacher understanding of the pedagogical practices needed to effectively guide students (Hubber et al. 2010). As noted above, Berland and Hammer's (2012) analysis suggests that individuals' social expectations (positional framings) interact significantly with the epistemological expectations (framings). From a framing perspective this suggests potential value in exploring the relationship between teachers' framing of their role and the nature of science knowledge (including its construction practices). Roberts (1996) offers useful insights into how teacher pedagogical understandings or "styles" affect students' opportunities and/or capacities to be active participants in construction and use of representations in any subject domain. In the current study these styles were considered as indicative of the positional, epistemological and conceptual framings of teachers. As noted above such framings will have effects on classroom practice and student understanding. For example, a teacher might assume that "authorised" sources such as textbooks sufficiently guide student understanding of the representations therein. However, as Bétrancourt et al. (2012) note in their study of science textbook graphical representations there is rarely instruction within such textbooks on how to "read" the representations and explicit support is needed for correct interpretation. This doesn't even consider student understanding of how such representations may have been originally generated nor how they function in the knowledge-producing discourse of Science (how science works). That is, with static textbook or other pre-resolved representations only their instrumental character is manifest but the constitutive function is not apparent (Magnani 2013). Potentially compounding student difficulties in learning chemistry concepts such as the particle model is the suggestion from researchers that not only

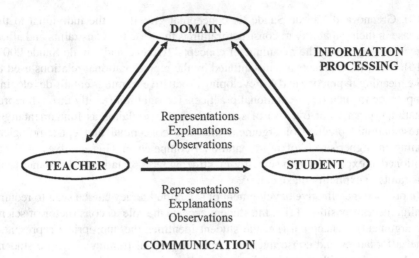

Fig. 11.1 Roberts' teacher in trialogue style

do teachers often not know how to succeed in developing such skills and under-standings but they may not even recognise the need for this instruction (Eilam 2012; Hubber et al. 2010). Roberts' ideal "trialogue" style is presented as concurring with teacher framings that recognise and address such needs.

In practice Roberts' (1996, p. 423) represents his ideal style as a three-way "tria-logue" with reciprocal linkages between each of teacher, student and domain. From a framing perspective the teacher frames their role as coach and negotiator of the meanings of domain representations and their refinement through a range of repre-sentational tasks. The arrow from teacher to student indicates the accepted wisdom of representations, as communicated by the teacher, while the reverse arrow indi-cates the students' prior or developing representations of the domain (Fig. 11.1). This interaction indicates the teacher does not assume the student's metacognitive ability (i.e. conceptual and epistemological framings) to recognise why authorized representations have been accepted as more compelling ways to establish knowl-edge in the domain. In this model, guided by some suitable scaffolding, students are encouraged to generate their own representations to explain observations and pre-dict future outcomes. They can then compare and reconcile these representations with those of their peers, and with those of their teacher, or those presented by their teacher as current within the science community. From a framing perspective the teachers frames their role as a guide in the recognition of representations' key fea-tures and subsequently how these features act as knowledge "justifiers" or "defin-ers" in the domain. In Roberts' words this constitutes no less than a 'precious metacognitive lesson' (Roberts 1996, p. 427).

Method

The case study took place in a grade 5/6 composite-class in a small school situated in a small semi-rural town in regional Victoria, Australia. In this education region students have each received a laptop computer as part of a government funded program. This provided one-to-one computer access for the purposes of enriching representational resources through the facilitation of animation and narration of explanations. The participating teacher was identified on the basis of her active engagement in previous education department professional development activities focused on using ICT effectively in the primary setting. She was an experienced, enthusiastic and able teacher, interested in developing her science teaching skills and knowledge, including the innovative use of strategies based on the development of students' representations using ICT.

The program consisted of two eight-lesson sequences where students were guided through a series of representational challenges focused around application of the particle model to change of state (sequence 1) and dissolving (sequence 2). Representational resources included computer stop-motion animation, traditional pen and paper, and hands-on materials such as beads for exploration and discussion. Data collection to inform this case study was mixed-method with a combination of quantitative and qualitative data collected and analyzed. The representational artifacts analyzed reflect a range of potential representational challenges applicable to the school classroom setting. Some of these were obvious assessment activities such a pencil and paper test (in which student explained a change such as the dropping water level of a fish tank), whereas others were effectively embedded, being for the most part indistinguishable from general learning activities, such as the animations and spontaneous representations produced by students over the course of the intervention. Teacher P was interviewed both prior to and after two sequences of approximately eight lessons of 70 min each. Other comments made in discussion were also recorded and used in the analysis. Six students were interviewed at a time between the two sequences. Teachers P identified these students as a cross-section representative of the range of abilities in each class. Considerations in the choice of students included literacy and numeracy levels and perceived general intelligence. The initial aim of these interviews was to explore student and teacher perceptions of the representational focus and tasks including development of understanding to capture critical influences on learning.

Data analysis followed a process of coding that was iterative or cyclical whereby the data was read and re-read to identify, corroborate or discard, and so reinforce or dismiss potential codes and themes. Allowing for rediscovery or reinterpretation this was continued until the researcher recognized the point of 'theoretical saturation' p. 74 (Phillips and Hardy 2002) – the situation where the identifiable in the data was completely instantiated by identified codes and resultant themes. Subsequently these resilient codes were further interpreted and grouped following the "learning as framing" theoretical literature. Positive development within these literatures is recognised as the construction of productive framings by students and teachers.

Findings: Teacher and Student Framings

Analyses of student and teacher interview transcripts and student work artifacts revealed patterns of respective re-framings as an effect of a representational focus utilizing ICT on middle-years science students' resources to learn scientific ideas of matter and its behaviour. This suggested that a framing perspective has promise as an indicator of the potential effects of a representation-focussed pedagogy in the pursuit of science literacy goals. This data is noteworthy in that student statements suggest that the representation-focused approach also facilitates the adoption of enactive (Son and Goldstone 2009) positional framings with regards to their role as constructors using the particle model for explanation of phenomena. There were also indications of the development of conceptual and epistemological framings reflecting the nature of science knowledge and practice. The key themes emerging were that this approach was engaging for a range of learners and was an effective vehicle for developing positive positional framings potentially productive of developing representational competence and conceptual understanding of the particle model.

Teacher P Framing Analysis

Teacher P Framing Practice as Science Abdication/Novice or Imposition

In the interviews Teacher P reported having taught science-focussed units once or twice a year over much of her 30-year career but over the past few years science learning was instead integrated within units that may include science knowledge. P attributed this transition to increasing assessment and reporting pressures – with science content needing to be opportunistically situated within themed literacy- and numeracy- focussed integrated units. In P's description of these integrated units students are, while guided generally in the development of questions, left to independently discover science domain knowledge and identify, interpret or construct explanatory representations. She notes, "you can't quite know what they will pick and so be on top of it". In this approach P may question students about their understandings, however insufficient knowledge in that aspect of the domain or time in a crowded unit curtails the coaching of students to link and resolve their representations with "authorised" ones. With respect to the knowledge constraint P effectively frames her role similarly to Roberts' Domain Novice (Roberts 1996) where, due to the breadth of possible science knowledge representations, she can provide limited guidance on, or specific assessment of, student understanding or use of representations, their conventions or explanations. The time constraint has P framing her role more like Roberts' Abdication model but the implications for student learning are the same in that her ability to provide domain guidance is effectively curtailed.

In reflection Teacher P noted the transmissive nature of her own school science experiences where "learning was what was in the text or what the teacher said was this or that". She acknowledged that this pedagogy had resulted in her framing her relationship with science as one where she would not question knowledge or challenge veracity of a text but rather "take what you are told rather than really talk about it and question it". This influence was evident in her description of planning for science-focussed units. When planning a focused science unit like "chemical science" in the past P referred to her available resources to check what the "experts say the key question should be, the key focus points, the key language to be used and that sort of thing – that becomes more like a particularly science-focused unit". Implicit in this discussion was the sense of "locking down" the information to be certain of its correctness, application and connection. In an attempt to guarantee accuracy of transmission to students, P's framing of her role resembles in style Roberts' Teacher Imposition with teacher as controller of the "accepted topic wisdom" through the demonstration of representations and explanations as resolved understandings. Her students' right to seek teacher clarification of understanding was evident, but (as with her own school experience) P did not describe student opportunities to construct and judge such representations by comparison with their own or peer efforts. In a sense the "legitimacy" of students' understanding through their need to link and resolve prior representations with "authorised" ones (the 'domain' in Roberts' terms) is effectively minimized. The students are presented with the "authorised" version as resolved conclusions rather than as discourse tools.

Representation Focus Facilitated Teacher P Reframing Practice as Trialogic

In the post-sequence interview P's comments revealed a new framing of her science-teaching role to one corresponding with Roberts' "Trialogic" style. In her discussion of her role she now appeared to frame herself as a coach and negotiator of representational meanings in a process of "assessment as they are learning". She noted as students constructed their representations, "dropping in questions 'Why have you got that like that?' or 'what's your intention with that?' or 'have you thought...?' or 'is this the best way you could do it?' – prompting their thinking in that. So that's learning and them refining at the same time, which I think is a valuable form of assessment isn't it! It's a process that's active – you don't just find out at the end what's happened." This emphasis on the "to and fro" between teacher and students regarding the domain marks a significant shift in P's framing of her role to one like Roberts' ideal "teacher in Trialogue" style. Through the experience of this study P legitimated the student construction of representations as a key part of student learning. She notes that she considers it "way more important" "for them to have to really think about that and to try to come up with their own understanding of it and be able to make a model to show that" "than just having it told to them".

However, P did not assume that students have the metacognitive ability to spontaneously recognize pertinent features of the representations. Rather she reiterated a responsibility to guide students regarding reasons behind the acceptance of representational modes, forms, conventions and interpretation. In this account her framing of her role "I suppose at some point if they don't actually come to the (accepted Science) understanding themselves through their own experimentation maybe they have to be told or shown or whatever". In her recognition and valuing of both the "representational exchange" between teacher and student and her role as a guide regarding the domain P exhibits a positional framing aligned to Roberts' ideal "Trialogue" style. P's positional reframing was further evident in her accounts of changes to her subsequent approach to teaching in other domains. For example, by the final interview P had also explored connections between a representational-focus and improved learning outcomes in the mathematics domain.

Teacher P Epistemological Re-framing

Comments by P indicated a changing epistemological framing that better aligns with contemporary literature viewpoints of the nature of science and science literacy regarding her choice of topic, planning and delivery for science-focused units. For example, P's apparent framing of scientific knowledge production had developed over the course of the two sequences to one of model-based discourse about ideas. In the first interview P described as a "real dilemma" her need to feel totally clear "in her head" about this expert knowledge including its real world connections. Without achieving this clarity she questioned her ability or authority to "get children to understand these things". Here, P's notion of scientific knowledge could be perceived as something utterly resolved to be transmitted for learning. Thus the great importance she places on her being "really knowledgeable in science" with the assumption that students presented with a resolved scientific concept will accept it logically and recognise its utility. After the two sequences, rather than seeing learning science as receiving and conveying a set of resolved factual knowledge, P perceived instead a process of actively applying and assessing explanatory models. Here canonical scientific knowledge now appears to be framed as a springboard for discourse as ideas are enacted or explored via representation and evaluated against observation as students are "actually trying to produce a model to show something." P elaborated with comparison to the class text 'My Side of the Mountain' where the main character was "doing things to find out". P noted he was learning best by "not relying (without question) on what scientists had said" but rather "relying on his study of that work". In this way students, "actually be being a scientist" not by hoarding "facts" but instead exploring concepts as they "played around" with ideas in representational discourse.

Student Positional Re-framing

Indications of Enactive Positional Framings

In the interviews following the first intervention many comments from different students indicated their development of enactive positional framings regarding their role as constructors and/or creators of explanatory representations. Their perception of the representational tasks – particularly the animation task – was as authentic and open challenges for them to apply and develop their understanding of the particle model. This active perspective was exemplified by male students B and G as they considered themselves as putting the particle model to use and thus projecting themselves as explainers (Greeno and MacWhinney 2006). For example, having been asked to choose what helped their learning in the sequence of lessons G responded, "I found just actually learning about it all, learning and then trying to put it into action almost" and B agreed. In this and following responses they position themselves as actors in the situation of explaining (Greeno 2009) in contrast to their previous experience (framing) of "sitting on the ground and listening" (B) where they "just learn about it then we don't actually use it that often" (G). In relation to their construction of their explanation, however, they used the expressions, "try to do it ourselves" (B), "actually do it" (G), "to actually get out there and do it" (B), "actually use it" (G) and "putting it (particle model) into action" (G). Other students also clearly positioned themselves as actors rather than observers with regards to the application of the particle model in the construction of explanations. For example, student K also described passive learning conditions as "boring" compared to the representation construction where "you are really involved with it and making it (the explanation)". In her comparison of passively receiving information to constructing it, K's partner D also revealed an enactive framing. She noted "also what we found on the internet was really confusing like the words and stuff they used and the way they put it whereas the way we put it makes it easier to understand". Student I noted that through creating the animated explanation she was able to include "explanation about why" rather than just the "what" responses of a multiple choice test. In the former she created a "fatter answer instead of a skinnier answer" and thereby she "found out more than you had before". H "liked having my own project to explain" and N described the task as "more of an activity to show your theory". Such offerings by students are indicative of them adopting an enactive positional framing with regards to constructing scientific explanations which has been considered a positive factor for learning (Son and Goldstone 2009).

Students Link Enactive Role with Meaning-Making Opportunities

Student comments further reveal that they associated their enactive role in the representation of ideas with successful learning outcomes. These clearly link "using" the ideas to learning through the creation of meaning making opportunities both for new or specialist meanings of individual terms and recognising the utility of the broader concepts. For example, students J and A discussed the development of their understanding of the key term, vibrating, through the construction of their animation of a cloud forming in a bottle. The idea of the vibration of particles in proportion to heat energy is core to understanding changes of state like the formation of a cloud. It was clear that the critical point for learning was the introduction to and re-representation of these terms by students in the context of constructing their explanation. J noted of the construction process, "at first we used little words when we didn't have our talk with you. We used little words like "move" and "shoved" and all that and then when we got (A: more interested and more into it) yeah then we used 'vibrating'." The researcher asked if they thought that was because the word "vibration" was discussed as J and A moved their "particles (circles) already on the screen and you could see what it might mean". J confirmed, "yes – vibrating in as if there is a particle there and it moves – it keeps moving and then we started understanding it as we made our diagram on the computer." Student M also noted constructing the explanation to "show the particles and how they vibrate" was "more effective" for his understanding of the meaning of the term in application. Similarly, student K also identified the act of re-representing the ideas allowed her to "get more stuff" because she was able to "actually see" the meanings of terms. Students G, B also link development of their understanding to their enactment of the meanings of words through the act of re-representation. Leading on from the researcher's comment that teacher P had first presented "the particle model way of thinking" G continued, "and then like having to try to put it into words, like your own words and just trying to *do* that stuff". B expanded "and (us) explaining how the particles move – I found that good, very useful". G noted in reference to constructing their animated explanation, "just listening about it you don't quite get the full understanding and then actually almost seeing it happen just helps". K noted, "It helped me understand it", later adding, "I'm not sure but I just get more stuff when I actually see…" and "I got it more as we went on (througmaking the animations)". K reasoned that this was because the student is "really involved with it and making it – then you think". D similarly reasoned, "Probably because you actually have to – you get to actually think about it and understand it before you can make it, before you can put it together so like you learn more because you have to think about it."

Student Conceptual Framing Developments

Students K and D

In each of the two sequences student pairs were expected to produce an animated particle-based explanation of a phenomenon. Grade 6 students, K and D worked consistently as a pair over the two sequences. They demonstrated established skills with the use of PowerPoint software and were able to represent their understandings readily to produce multiple PowerPoint versions. K and D generally worked well within the sessions and between them produced two versions for each of the anima-tion tasks. The four paired animation and two individual paper-test artifacts pre-sented span the course of the classroom sequences including other representational activities such as the group bead explanations. The extra draft versions each dis-played variation from the final versions and so provided additional insight into their framing developments. The understandings represented therein show progression by K and D in sophistication of use of particles as an explanatory basis for changes of state. These multiple representations reveal a pathway including apparent ideas of "macro" particles, particles with macro properties (particles themselves dissolv-ing) to en masse sub-micro particle behaviour. Additionally, however, these provide interesting insight into the development of conceptual framings as evidenced by the choice of arrangements in their PowerPoint animations. Over the course of the series of representations K and D can be seen to progressively foreground particle aspects of their explanations and draw on productive constraints. The inclusion of the student work below is to explore this second development – that is, the potential of student artifacts to indicate educationally productive changes to the conceptual framings of learners.

Framing Analysis of K and D's Representational Artefacts and Related Interview Comments

In these early attempts particle explanations are not foregrounded but enmeshed within other aspects of the empirical phenomena. K and D's first explanation (Fig. 11.2) has as its basis the movement of water from the ground to the plant leaves rather than a change of state involving the surrounding air. The blue circles are labelled as 'particles' and the text notes, 'the plant is colder than the surrounding air. It has to be this way for frost to form'. The sense of "particles" is taken to be as macroscopic liquid droplets moving through the soil and the air temperature as related to drawing these up the outside of the plant. Their second explanation of frost, though ostensibly more scientifically accurate with its attribution of the source of water to the air, again does not particularly foreground the particle basis of the explanation. The accompanying text states 'when the air gets to a certain tempera-ture the particles come *from* the air and form on the plant as frost' and 'when they *actually form* on the plant *they become* the frost' (my emphasis). This suggests they are beginning to integrate (verbally at least) ideas of component particle entities continuous across changes of state. The use of iconic representations such as drips for liquid water particles and corresponding text such as 'the white stars represent particles in the form of ice/snow' indicate that the girls were still struggling to con-ceive of or at least appropriately represent such sub-microscopic entities. This is also evident in the lack of evaporated particles in latter slides that note evaporation back 'to the clouds where it initially started'. Possibly drawing from their knowl-edge of water cycles D explained the inclusion of the clouds as it "makes more sense – the full cycle of what happens". However, K and D only represent the move-ment of water somewhat by proxy in the form of arrows rather than particle

Fig. 11.2 Selected slides from K and D's first and revised (second) versions of an animated explanation of the occurrence of frost

representations. These aspects are taken as indicative that these students are not yet framing the particulate view of the phenomena as the means of conjecture.

Paper-Based Assessments

By the time of the paper test at the end of the first sequence both K and D independently depict water particles in the gas phase to represent evaporation over time of water from a fish tank (Figs. 11.3, 11.4 and 11.5). The implicit narrative of a water cycle is still obvious in both examples with the depiction of destination clouds but particles are now represented between clouds and liquid and are spaced relatively widely as per gases. K labels the particles as 'water' both inside and outside the cloud and D notes 'they become a gas' in her text. D was one of the few students who also included a sound causal mechanism for the evaporation at sub-boiling temperatures with arrows to represent particle movement in the liquid (collision and

Fig. 11.3 Detail of Student K's explanation of evaporation over time from a fish tank showing exaggerated water particles between the source and clouds

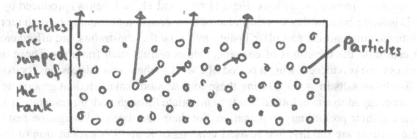

Fig. 11.4 Detail of Student D's explanation of evaporation over time with idealised fish tank and exaggerated particles used to represent a mechanism for evaporation. The accompanying text stated, 'Particles move around more when they are warmer. They bump off each other and they occasionally will be bumped out of the water. They become a gas'

Fig. 11.5 Detail of Student D's explanation of evaporation over time with idealised cloud and exaggerated particles showing particles outside of a cloud. The accompanying text stated, 'Eventually, particles out of the tank will make it outside and will become a cloud'

ejection). Significantly the particulate aspects of the explanation are now more prominent in the explanation of the phenomena and are foregrounded through relative size (Fig. 11.3) or minimizing of other detail in the overall representation (Fig. 11.4). These appear as purposeful over- and under-representations of aspects of the empirical phenomena. That is, aspects of information in the situation have been attended to differently for the purposes of conjecture. While these may seem indications of traditional conceptual growth through recognition of particles in the gas phase, at the point of the interview some weeks later both K and D still continued to harbour a degree of uncertainty around gaseous water. This was revealed by their responses to the researcher's question, "can it (water) be somewhere between the clouds and the ground?" K: (pause) yeah probably, but wouldn't it fall to the ground and then evaporate? R: Yes it can – so where is it when it evaporates? K: the clouds D: It goes into the sky. Such distortions in the paper-based assessment may thus be better viewed as further development of conceptual framings regarding the nature and means of science argument.

Second Sequence Artifacts

The second sequence animations (Figs. 11.6a–c and 11.7a, b below) produced by K and D indicate both further development of their conceptual framing of such representational conventions and offer insight into how the constraints and affordances that constitute the meanings of concepts help construct such framings. These two versions were produced during a period of 2 weeks where several sessions focussed on dissolving salts in water. During these four sessions small student groups were also encouraged to use coloured beads to negotiate through verbal presentation and enactment their preliminary explanations for their dissolving investigation results. These artifacts are the first and revised versions of K and D's explanation of their observation of differing dissolving rates dependent on water temperature.

In these explanations particle representations are significantly removed from accounts of the empirical observations (depictions of the investigation vessel and descriptions of the dissolving salt) by their placement on separate slides. Both ver-

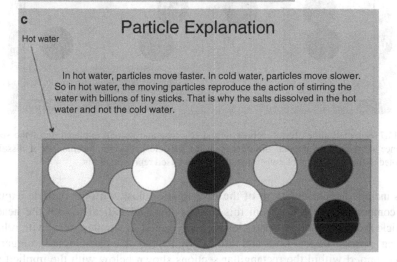

a   what happened with cold water?

1. We added the Sodium Carbonate first and it
   clouded the water. The Sodium Carbonate
   did not dissolve, neither did the Sodium
   Bicarbonate. Instead of dissolving it came up
   like trees. It formed a hard disk when we
   tried to rinse

b   what happened in hot water

- Both substances dissolved. It made a popping
  noise and it went cloudy. From the bottom to
  the top it cleared. The temperature was 39 1/2
  degrees Celsius. There was a lot of bubbles.

c   Particle Explanation

Hot water

In hot water, particles move faster. In cold water, particles move slower.
So in hot water, the moving particles reproduce the action of stirring the
water with billions of tiny sticks. That is why the salts dissolved in the hot
water and not the cold water.

Fig. 11.6 Selected slides of K and D's first explanation of the second sequence displaying segregation and foregrounding of particle representations

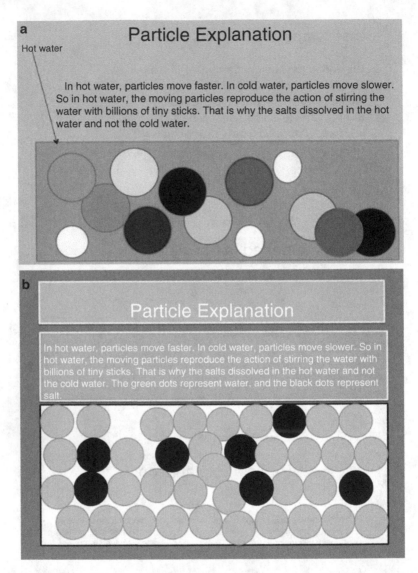

Fig. 11.7 Selected slides from each of K and D's first and second explanation of the second sequence showing fundamental differences in the particle behaviour mechanism of dissolving attributed to the translation of constraints from bead-based representations

sions included representations of the investigation flask with the particle explanation component separated from this macro-view and indicated with the heading 'Particle Explanation'. The particles are represented as simple circles with relative spacing appropriate for solids and liquids. Stylistically the particles are exaggerated and contained within the rectangular sections shown below with the implicit suggestion that this idealised space is within, but independent of, the container. Both first and revised explanations include the text, 'In hot water, particles move faster. In cold water, particles move slower. So in hot water, the moving particles reproduce the action of stirring the water with billions of tiny sticks. That is why the salts dissolved in the hot water and not the cold water.' The removal and exaggeration of the particle diagrams along with this focused mechanistic verbal conjecture clearly focus the reader on the particles as central to the explanation. This is seen as indicative of conceptual framings regarding science argument whose distortions afford and constitute focus on specific aspects of the empirical phenomena (cf Woods and Rosales 2010).

The most significant difference between these versions of the second sequence is evident in K and D's diagrammatic representations of particles. In the first explanation (Fig. 11.7a) the salt particles are clearly themselves shrinking (cf Fig. 11.6c) whereas in the second (Fig. 11.7b) the particles are shown separated but are themselves unchanged. This fundamental development in the presented causal mechanism for dissolving between these two accounts corresponded with the bead-based negotiations of the same period. That is, K and D have apparently recognised and employed the constraints and affordances inherent in the bead representations to further develop their conceptual framings and reconstitute the meanings of particle concepts in this context.

Discussion

The analysis above suggests pedagogical practices that prescribe a series of representation constructions by students are conducive to the co-development of productive framings by students and teachers. While the positional, conceptual and epistemological framings explored here are not viewed as separately variable (van de Sande and Greeno 2012) these distinctions proved valuable for identification and analysis of the sorts of frames students and teachers engage in, and as a result of, this pedagogical approach. These framings, of course, mutually interact in the development of individual and group understanding of scientific knowledge construction. In describing her actions in this teaching and learning approach, Teacher P framed her role like Roberts' ideal Trialogic style whereby students were 'entitled and expected to question, challenge, and adapt concepts and methods of the subject-matter domain for purposes of their activity and understanding' (Greeno and van de Sande 2007, p. 9). Greeno and van de Sande (2007) describe such practice as entailing a vestment of conceptual agency for teacher and students. This sense of vestment was evident in student interviews where students appeared to have adopted an enactive positional framing (MacWhinney 2005) as they perceived themselves to be "really involved with it and making it (the explanation)". Practices that promote such enactive perspectives entail 'a deeper, more embodied level of processing' (MacWhinney 2005, p. 199) and more significant opportunities for conceptual growth (Greeno and van de Sande 2007). Student statements reflected this connection as they noted the approach enabled them to explore and refine possible meanings through guided representation construction. The student artifacts also presented evidence of a refinement of conceptual framings through attunements to conceptual constraints and affordances (Greeno and van de Sande 2007). Their progressive foregrounding of the scientific explanation over the course of the sequence of representational challenges can be seen as development of conceptual framings in terms of employing deliberate distortions in their organisation of different aspects of the information for clarity of explanation. Though not explored here for reasons of parsimony such purposeful idealisations could also be indicative of epistemological framings associated with student understandings of what or how we come to know in Science.

While the discussion above explores the interconnectedness of the three framings described their recognition and analysis provides a multi-frame basis for clarifying learning gained from representation-focussed pedagogy. Bing and Redish (2012) note that even (scientifically) inaccurate student responses may show inherent learning value from a different framing viewpoint. Both Justi et al. (2009) and Linn et al. (2010) suggest that representation generation activities increase student awareness of the complexity of phenomena and challenge illusions of full understanding. Rather than hindrances to learning Linn et al. suggest these could 'encourage learners to revisit their existing knowledge' (Linn et al. 2010, p. 242) to develop a more sophisticated understanding of such representations and ask more effective questions. Such questions could be associated with developing conceptual and epistemological framings of science knowledge production. As one student in this study offered regarding why the teacher had students construct particle model explanations, "it was more of an activity to show your theory" with the reasoning "because then you can understand why and why not it's right". A multi-frame analysis of learning gains brings a different focus on suitable representational challenges including what, how and when authorised representational conventions are best introduced for developing associated understandings.

References

Bateson, G. (1972). *Steps to an ecology of mind*. San Francisco: Chandler Press.

Berland, L. K., & Hammer, D. (2012). Framing for scientific argumentation. *Journal of Research in Science Teaching, 49*(1), 68–94.

Bétrancourt, M., Ainsworth, S., de Vries, E., Boucheix, J. M., & Lowe, R. K. (2012). Graphicacy: Do readers of science textbooks need it? In E. de Vries & K. Scheiter (Eds.), *Proceedings EARLI special interest group text and graphics: Staging knowledge and experience: How to take advantage of representational technologies in education and training?* (pp. 37–39). Grenoble: Université Pierre-Mendès-France.

Bing, T. J., & Redish, E. F. (2012). Epistemic complexity and the journeyman-expert transition. *Physical Review Special Topics: Physics Education Research, 8*(1), 010105, 1–12.

Eilam, B. (2012). *Teaching, learning, and visual literacy: The dual role of visual representation in the teaching profession*. New York: Cambridge University Press.

Elby, A., & Hammer, D. (2010). Epistemological resources and framing: A cognitive framework for helping teachers interpret and respond to their students' epistemologies. In L. D. Bendixen

& F. C. Feucht (Eds.), *Personal epistemology in the classroom: Theory, research, and implications for practice* (pp. 409–434). Cambridge: Cambridge University Press.

Ford, M., & Forman, E. A. (2006). Refining disciplinary learning in classroom contexts. *Review of Research in Education, 30*, 1–33.

Goffman, E. (1974). *Frame analysis: An essay on the organization of experience.* Cambridge, MA: Harvard University Press.

Greeno, J. G. (2009). A theory bite on contextualizing, framing, and positioning: A companion to son and goldstone. *Cognition and Instruction, 27*(3), 269–275.

Greeno, J. G., & MacWhinney, B. (2006). *Learning as perspective taking: Conceptual alignment in the classroom.* Paper presented at the 7th international conference on learning sciences, Bloomington, 27 June–1 July 2006 (Poster).

Greeno, J. G., & van de Sande, C. (2007). Perspectival understanding of conceptions and conceptual growth in interaction. *Educational Psychologist, 42*(1), 9–23.

Hammer, D., Elby, A., Scherr, R. E., & Redish, E. F. (2005). Resources, framing, and transfer. In J. P. Mestre (Ed.), *Transfer of learning from a modern multidisciplinary perspective* (pp. 89–120). Greenwich: Information Age Publishing.

Hofer, B. K. (2001). Personal epistemology research: Implications for learning and teaching. *Educational Psychology Review, 13*(4), 353–383.

Hubber, P., Tytler, R., & Haslam, F. (2010). Teaching and learning about force with a representational focus: Pedagogy and teacher change. *Research in Science Education, 40*(1), 5–28.

Justi, R., Gilbert, J. K., & Ferreira, F. M. (2009). The application of a 'model of modelling' to illustrate the importance of metavisualisation in respect to the three types of representation. In J. K. Gilbert & D. F. Treagust (Eds.), *Multiple representations in chemical education* (pp. 285–307). Dordrecht: Springer.

Linn, M. C., Chang, H. Y., Chiu, J. L., Zhang, H., & McElhaney, K. (2010). Can desirable difficulties overcome deceptive clarity in scientific visualizations? In A. Benjamin (Ed.), *Successful remembering and successful forgetting: A festschrift in honor of Robert A. Bjork* (pp. 239–262). New York: Routledge.

MacWhinney, B. (2005). The emergence of grammar from perspective taking. In D. Pecher & R. Zwann (Eds.), *The grounding of cognition* (pp. 198–233). Cambridge, UK: Cambridge University Press.

Magnani, L. (2013). Is abduction ignorance-preserving? Conventions, models and fictions in science. *Logic Journal of IGPL, 21*, 882.

Phillips, N., & Hardy, C. (2002). *Understanding discourse analysis.* Thousand Oaks: Sage.

Prain, V., & Tytler, R. (2012). Learning through constructing representations in science: A framework of representational construction affordances. *International Journal of Science Education, 34*(17), 2751–2773.

Redish, E. F. (2004). *A theoretical framework for physics education research: Modeling student thinking.* Paper presented at the proceedings of the Enrico Fermi Summer School, Course CLVI, Bologna.

Roberts, D. (1996). Epistemic authority for teacher knowledge: The potential role of teacher communities: a response to Robert Orton. *Curriculum Inquiry, 26*, 417–431.

Scherr, R. E., & Hammer, D. (2009). Student behavior and epistemological framing: Examples from collaborative active-learning activities in physics. *Cognition and Instruction, 27*, 147–174.

Shemwell, J. T., & Furtak, E. M. (2010). Science classroom discussion as scientific argumentation: A study of conceptually rich (and poor) student talk. *Educational Assessment, 15*(3), 222–250.

Son, J. Y., & Goldstone, R. L. (2009). Contextualization in perspective. *Cognition and Instruction, 27*, 51–89.

van de Sande, C., & Greeno, J. G. (2010). *A framing of instructional explanations: Let us explain with you instructional explanations in the disciplines* (Vol. 2, pp. 69–82). Boston: Springer.

van de Sande, C., & Greeno, J. (2012). Achieving alignment of perspectival framings in problem-solving discourse. *The Journal of the Learning Sciences, 21*, 1–44.

Vosniadou, S., Vamvakoussi, X., & Skopeliti, I. (2008). The framework theory approach to the problem of conceptual change. In *International handbook of research on conceptual change* (pp. 3–34). New York: Routledge.

Weisberg, M. (2007). Who is a modeler? *British Journal for the Philosophy of Science, 58*, 207–233.

Woods, J., & Rosales, A. (2010). Virtuous distortion: Abstraction and idealisation in model-based science. In L. Magnani, W. Carnielli, & C. Pizzi (Eds.), *Model-based reasoning in science and technology* (pp. 3–30). Berlin: Springer.

Chapter 12
Emerging Developments and Future Questions

Mark McDermott

As is emphasized repeatedly throughout this book, the language of science is multimodal. The discipline of science relies on the use and integration of multiple modes of representation to both develop understanding of natural phenomenon and to communicate that understanding. As Yore et al. (2003) assert, language is involved in both the communication about and the "doing" of science. In a related way, multimodal language can play both a communicative role and a knowledge generating role in the science classroom. Realizing this, science educators have begun to focus on the use of the multimodal language of science as more than just a way for students to communicate previously developed knowledge (Gunel et al. 2006). The contributing authors for this book, although varied in their approaches, recognize the potential power of utilizing communication associated with the multimodal language of science to serve as an epistemological tool available to students as a way to develop their scientific literacy. These authors generally agree on the need to improve the science learning as well as the motivation and interest of students across the globe. In addition, their work has at least begun to explore the role focusing on multimodal language practices as a means for facilitating this improvement. However, the specific pedagogical approaches employed and the success of the pedagogical approaches relative to stated goals have varied across settings.

Although there is obvious divergence in terms of the specifics of the theoretical frameworks guiding different studies presented in this book, there is a general trend toward what Klein (2006) discussed as second generation views on cognition. Rather than taking the first generation position that the use of appropriate scientific language follows thinking about scientific ideas as a means for reporting information, the work here aligns with a position that language can be used to shape ideas about science (Klein 2006). As students engage in the use of language to describe scientific ideas, connections between related aspects of the targeted concepts are

M. McDermott (✉)
University of Iowa, Iowa City, IA, USA
e-mail: mark-a-mcdermott@uiowa.edu

© Springer International Publishing Switzerland 2016
B. Hand et al. (eds.), *Using Multimodal Representations to Support Learning in the Science Classroom*, DOI 10.1007/978-3-319-16450-2_12

strengthened (Yore and Treagust 2006). As Yore and Treagust (2006) claim, adoption of this second generation perspective can lead to a more "pragmatic" view of logic or reasoning related to a "flexible mixture of natural and disciplinary language" (p. 209). This perspective seemingly aligns with the argument of Anderberg et al. (2008) reported by Prain and Hand in the introduction chapter of this book about the "dynamic and ambiguous character of the relations between students' meanings, conceptions, and expressions". Anderberg et al. (2008) suggest a position likely supported by the authors of this volume that student reflection on the adequacy of the links between meanings, conceptions, and different expressions can lead to language becoming a tool for learning. In this way, effective engagement of students in multimodal learning environments can potentially provide a means for enacting a second generation view of cognition that allows students to utilize language, as well as experiences, to move from "fuzzy" conceptions of scientific ideas to more disciplinary-accepted explanations.

The specific pedagogical interventions employed in the studies reported here enact different theoretical frameworks as described by the authors of each chapter. However, a sense of convergence exists relative to the overarching goal of instruction which could be summarized as an attempt to create learning environments in which students:

1. develop a personal understanding of targeted science concepts that aligns with the currently accepted views of the scientific community.
2. became aware of the processes involved in the development of scientific knowledge in the scientific community.
3. have an opportunity to participate in those processes as they develop personal scientific understanding.
4. come to understand that it is part of their role as a learner to contribute to the development of understanding.

Students will ideally come to realize and utilize the simultaneous power of learning from exploration of multiple representations of targeted phenomenon, ideas, and concepts, as they effectively communicate about the natural and physical environment in a multimodal way. Prain and Hand's description in the introductory chapter of *writing* as "a key epistemological tool for learning, in that drafting and revising processes enabled students to build and review links between classroom activities, conceptual understandings, and their expression" can be extended to the use of multimodal communication. The studies reported here, when taken together, suggest that to gain this realization, both the knowledge generation and communicative aspect of multimodal language must be continually encountered and encouraged in classroom settings. One aspect then of developing a true scientific literacy includes the development of a "multimodal competency". This multimodal competency includes both an awareness of the potential for representing one idea or concept in multiple ways as well as an awareness of the strategies needed to effectively link representations of different but related concepts (including text) in an overall piece of communication (McDermott and Hand 2013). Ideally, simultaneous with the development of this multimodal competency is development of an awareness that using these representational tools affords individuals an opportunity to better understand the natural and physical environment and to clearly communicate that understanding to others. Finally, students engaging in the use of multimodal language as an epistemological tool would also hopefully come to realize their role as a constructor of knowledge rather than simply a receiver of already determined knowledge from others, gaining the "enactive positional framing" (MacWhinney 2005) that Carolan describes in his chapter.

Students, however, are not the only participants in the learning environment. Teachers, too, must come to understand the role students can play in the construction of knowledge and the implications student construction of knowledge has for their work in setting up an environment in which knowledge construction can most effectively take place (Hubber et al. 2010). If the use of multiple representations and multimodal communication are seen as effective tools for encouraging this knowledge construction, then research efforts must also be focused on the teaching processes involved in encouraging and nurturing their effective use in classroom settings. Research must not only seek to clarify the cognitive activity and outcomes of students as they engage with multiple representations and use of multimodal communication, but also the teaching practices that will more likely lead to this engagement resulting in positive conceptual gains. This clarification obviously requires a broad based, general understanding of learning with multiple representations and multimodal communication, as well as an understanding of how it manifests itself within the particular disciplinary characteristics of learning about science.

Many of the studies discussed in this book suggests that although there is a relationship between effectively utilizing multiple representations as a part of classroom practice and improved conceptual understanding, this benefit does not spontaneously develop through the simple inclusion of multiple representations in the learning environment or requesting students create multimodal artifacts or products. Purposeful planning to effectively and systematically embed multiple representations in instruction, as well as to develop tasks in which students are asked to develop multimodal communication is necessary (and even then not always sufficient) to develop learning situations that positively impact conceptual development. To better assure the consistency of the benefit of these sorts of interventions and tasks, the preceding chapters suggest further clarification of (at minimum) the following issues:

1. How do we appropriately differentiate the idea of multiple representations from the idea of multimodal communication in a manner that helps further develop effective classroom practices?
2. What are different ways to categorize the multiple representations that students use and how might this impact further study?
3. What is the most appropriate relationship between multiple representations, multimodal communication, canonical forms of communicating about science and alternative forms of communicating about science?
4. How is student reasoning related to the use of multimodal communication?

5. How do we best measure student multimodal competency so that it reflects our understanding of the traits of multimodal communication that are most beneficial in developing conceptual understanding?
6. What are the pedagogical implications for our answers to the preceding questions?

Multiple Representations vs. Multimodal Communication

A fundamental question, lending itself more to discussion and consensus building rather than understanding through research methods is the issue of clarifying the distinctions between and relationships among the ideas of "multiple representations" and "multimodal communication". Although in some literature these terms are treated as synonymous (McDermott 2009), the terms have unique meanings. Delineating specifically what each means may be more important in terms of clarity in reference to different types of tasks and pedagogical interventions, but could also set the stage for further research. Much of the literature and research that grounded early work with multiple representations and multimodal communication dealt with multimedia learning (Mayer 1996). This work tended to focus on exploring the impact of the use of multimedia learning environments which contained visual or auditory modes outsides of text on student understanding in a wide range of educational contexts. The main goal was to assess whether or not student exposure to different forms of information was beneficial in terms of developing understanding of targeted concepts. This literature and the research associated with it, including analysis of the cognitive load that this exposure to multimedia environments placed on students (Sweller et al. 1998), did not typically deal with students acting as constructors of their own representations of information in any mode. Rather the student was a receiver of information being presented in a multimodal format. In many cases, these multimodal environments were technology driven and outcomes related to student understanding were compared between unimodal and multimodal environments.

The work with multimedia environments provided some impetus for science educators to begin to consider the use of a variety of modes in science learning environments. As described in the introductory chapter, some researchers developed an interest in the use of multimodal representations from earlier work in the area of writing-to-learn activities in science. The chapters from Simon, from Tolppanen, Rantaniity, and Aksela, from Gunel, Kingir, and Aydemir, from Nam and Cho, and from McDermott and Hand, describe projects aligned with and to some degree emerging from this line of research. Their work has begun to focus on how students create written products in which modes other than text are infused. Often, the multimodal writing product is a communication piece that summarizes connections and relationships between several targeted concepts within an overall unit of study. As work in this area continues to refine both the sought after characteristics of the multimodal written products, the relationship between different

characteristics of multimodal products and conceptual development, and the pedagogical approaches most likely to lead to conceptual benefit, emphasis is still
squarely placed on student development of the written text, student determination of
the modes used in the communication and the strategies utilized to link modes
together, and assessment of the degree to which the engagement in this type of
activity leads to greater opportunity for students to clarify and enhance their
knowledge.

A similar avenue of inquiry grounding exploration of multiple representations
and multimodal communication is seen in other chapters in this volume. The work
reported by Carolan, Tytler and Hubber, and by Tang, Ho, and Putra related to representational challenges involves focus on utilizing a carefully crafted sequence of
student interaction with different representations (including textual, visual, and kinesthetic) as a means for scaffolding development of understanding of the multiple
aspects and characteristics of a somewhat more narrowly defined targeted concept.
However, students are not simply presented with different representations. Rather
they are engaged in the development of their own representations generated from
different classroom experiences. Comparison of different representations is seen as
a way to progressively develop a more nuanced and appropriate understanding of
the targeted concept. In many cases, teacher generated or presented representations
are compared to student generated representations as a means to evaluate the relative strengths and weaknesses of the different representations of scientific ideas, and
a way to further clarify understanding. Ideally, these sequences aid students in the
development of a more complete understanding of the targeted concept while at the
same time offering opportunities to explore the knowledge generation potential of
using different types of representations, explore the approximate nature of representations, and foster beneficial debate about both the representations utilized and the
concepts represented.

A cursory consideration of the products created as a result of the two seemingly
divergent pathways can lead both to a beneficial clarification of terms and a potentially dangerous overgeneralization. On one hand, in terms of clarifying the distinction between multiple representations and multimodal communication, the products
created in situations where modes outside of text are embedded in writing-to-learn
tasks can clearly be labeled "multimodal communication". These products demonstrate the use of different types of modes of representation within one communication piece. A student, in generating multimodal communication, may choose to
represent a specific concept or aspect of a concept with more than one mode, and
therefore utilize multiple representations. Often, concepts are at minimum represented in textual descriptions and with one other non-text mode. However, multiple
representations may not be a guaranteed outcome. In fact, a more integrated multimodal communication piece may be one in which although many different modalities are used, specific modes have been purposely selected as the sole way to
represent specific ideas. Conversely, the products developed in the chapters describing representational challenges are quite clearly multiple representations as they
involve using different modes to represent and re-represent a specific targeted concept. In fact, the term "re-representation" is often used to describe the process of

moving from one mode of representation to another. The constellation of representations that are generated by an individual student or a class as a result of participation in a sequence of representational challenges would be classified as "multimodal" in that they will include several different modes of representation. However, although any one specific representational product may include more than one mode (for example a picture that is linked to a text description), this is not a guaranteed outcome.

Focusing attention solely on the differences between the two approaches could lead to an unwarranted distinction and conclusion. For example, the differences could be construed as evidence that one group is only concerned with the effective linking of different modes representing different but related ideas in a broad piece of communication and not the specific characteristics of each representation. The other group could be construed as concerned only with re-representing a specific concept in different ways, with different modes, without concern for how different modes are linked. Although the different approaches do lead to different pedagogical interventions, they have some very important similarities. Both approaches are attempting to engage students in multiple "translations" in which prolonged, productive engagement with a concept or concepts is encouraged as students consider multiple ways to represent the concept *or* the linkages between different modes *or* both. Both approaches are based on the premise that these multiple translations can lead to student awareness of the inadequacy of their current understanding and drive students toward a clarification of that understanding. Both approaches seek to help students (and teachers) realize the power of representational tools as modes for communicating understanding to help further the social negotiation necessary to come to consensus about explanations about the natural and physical environment and for providing individuals a means for personal negotiation of meaning. As such, both representational activities have potential to achieve the previously discussed goal of representational use simultaneously fulfilling a communicative and knowledge generative role.

Although the comparison of these two approaches helps highlight the technical distinction between multiple representations and multimodal communication, a more important question emerges: How can activity infuse consideration of the effective aspects of both approaches? How can students who are engaged in creating multimodal writing tasks infused with multiple modes of representation and students engaged in representational challenges be encouraged to consider both the effective ways to link different modes together and the selection of the most appropriate means for representing an idea or thought from a number of modal possibilities. Should engagement in the creation of multiple representations of a single concept and the consideration of which mode is most accurate, effective, and thorough precede instruction aimed at students developing well integrated multimodal communication pieces? Or, should the engagement in discussion concerning which type of representation is most appropriate for a given circumstance be embedded in the development of more overarching multimodal communication pieces? Or, are there other possibilities that might be equally or more effective?

Regardless of the approach leading to the use of multiple representations and multimodal communication as a tool for learning, the unifying characteristic of the studies in this book is the emphasis on either simultaneous or progressive use of more than one mode of communication. In the sections that follow, we highlight questions emerging from a careful look at the modes utilized.

Questions Related to Categories of Modes of Representation

Linebarger and Norton-Meier, in their chapter, discuss a more global perspective of modal use related to early-learners making sense of their environments when they describe modes including aural (e.g., music), visual (e.g., pictures, art), aural/visual (e.g., television, video games), linguistic (e.g., storytelling), gestural (e.g., body control, emotion), and spatial (e.g., geographical, architectural). They then point out that even these very young learners can not only interpret the characteristics of the world around them through processing of these different modes, but they can also use the modes to further develop their understanding and communicate their understanding. This application of a categorization scheme and the related discussion provides a framework for thinking about categorizing modes used by older students.

Students engaged in the various tasks reported throughout this book created products that included text, diagrams, graphs, charts, mathematical equations, animations, and other modes. They also engaged in comparison between their representations and representations of these various modes created by others. These modes were created in multiple ways. Some were generated kinesthetically (through role-playing or manipulation of physical objects), others on paper (through handwritten or computer generated drawings, graphs, tables, etc.) and still others through technological means (through computer simulations). Regardless of the method of generation, the representations tended to fall into three broad categories.

1. Representations utilized as a tool to carry out a scientific process or aid in application of a scientific practice or skill. For example, Venn diagrams were utilized to aid in the process of classifying matter in Tytler and Hubber's chapter. Another example could be a student using a data table to organize observations collected in an experiment, the use of a graph to display trends, or the use of a mathematical equation to show a relationship.
2. Representations utilized as a tool to represent actual objects in nature. Diagrams were utilized in student products throughout the chapters as a way to represent actual structural components that exist in natural phenomenon. Using globes and balls to represent the Earth, moon, and sun is another example of this. Often, a non-text representation of this type will depict structures of natural phenomena that are too small (e.g. atoms or molecules that make up matter) or two large (planets) to be directly observed and manipulated in the classroom.

3. Representations utilized as a tool to represent a term or a process or an idea. Tytler and Hubber reported how students used role playing to help clarify the meaning of the term vibration. The representation (in this case students role playing different types of movement) was utilized to help pinpoint the appropriate term to apply to an action that could then be utilized to describe different natural phenomena.

The questions that arise from considering these different categories of modes include determining whether or not there are more beneficial sequences of use of modal types or more beneficial connections between different modal types. As students encounter new concepts, would engagement with a specific type of mode or utilization of a specific generative process be more beneficial before the others? Should students engage in use of all types of modes and all types of modal generating processes with all concepts or are certain concepts more appropriately paired with specific modal types or specific generation processes? Is it critical or beneficial that students are aware of the nuances that differentiate these different types of modes or generative processes or is it adequate that they utilize them without realizing the different categories of modal representations? Would the generative power of representational use be facilitated through student understanding of the particular type of mode that they are employing at any given time in a pedagogical sequence?

Questions Related to Canonical and Alternative Formats for Communicating About Science

One goal often expressed in the chapters of this book is to help students better understand the processes of science through engagement with multimodal language. On one level, students are encouraged to utilize multimodal communication because scientists themselves use representations of phenomenon both to better understand ideas they are attempting to clarify and to communicate information about these ideas to outside audiences. Sometimes, this process is an iterative one and the continual refinement of ideas and understanding is facilitated by the intermittent construction and communication of knowledge mediated through representational use. In engaging students in behavior that reflects what scientists do, we are not only attempting to help them develop their own conceptual understanding of the knowledge about nature that has been constructed by scientists in this manner, but also become more aware of the general characteristics of scientific inquiry. Two questions related to multimodal communication and this process emerge.

As described in the introductory chapter of this book, an ongoing debate in the literature regarding using written communication in science is the question of the relative value of students engaging in writing that is representative of the canonical or traditional writing styles of the discipline as opposed to writing that is more creative or alternative. Some would argue that if we are to help students become more aware of the disciplinary characteristics of science, this awareness must include experiences in which students engage in discipline specific communication, structured in disciplinary specific ways, utilizing appropriate and specific disciplinary vocabulary. Proponents of this view would argue that because scientists communicate their findings in a somewhat typical way, when students are to engage in writing as a part of their science classes, they should engage in creating similar written products. Others maintain that engaging students in alternative forms of writing, in which they are allowed to utilize more creative means of expression and in which they are allowed (and often encouraged through the process) to actually translate the language of science into a more approachable language for the audience written to, is more likely to lead to the writing being a means for facilitating knowledge construction as opposed to summarizing already developed knowledge.

In some of the chapters in this book (such as Tang et al.), the authors have argued that the use of multiple representations and engaging in multimodal communication is a means for developing disciplinary literacy that involves the refined use of appropriate disciplinary vocabulary. Others have offered projects (such as Simon) that involve more alternative writing formats in which students are encouraged to be creative in the written products that they infuse with multiple representations. Some instances even combine a less traditional summary writing task with exploration of canonical forms of scientific communication as a part of a lesson designed to encourage students to discover strategies scientists typically use to link different modalities used in communication together. To some extent, the issue of which of these types of writings should be encouraged is answered: scientists really have to employ both in their work, and both have been shown to provide students with conceptual benefit. One emerging question in this context is whether alternative or more traditional writing is more appropriate, and ultimately, more beneficial, when multiple modes of representations are infused. Could one type or the other be encouraged early in the process of conceptual development as opposed to later, or is one type of multimodal communication more beneficial in terms of student outcomes when used as a summary task or an engagement of prior knowledge, or as a means for concept development or other instructional purpose?

Possibly a more fundamental question is whether or not the role or the use of non-text representations are altered based on whether the representation is being used as a part of a more traditional, canonical communication or a more alternative or creative one. Should a mode outside of text reiterate or to some degree repeat information that has already been discussed in the text or should the mode outside of text represent information that is better communicated in a non-text fashion and is therefore presented as a non-text mode precisely because it cannot be adequately addressed with text? Does the best use of the mode change based on whether the author is composing a formal lab write up or a magazine article reporting about a science idea? If the latter is the more appropriate case, then development of a multimodal competency would involve developing skill in ascertaining what concepts or ideas are more accurate and more approachable when presented in a non-text way in addition to understanding and applying ways to effectively link different sections and different modes of the communication. In terms of classroom settings, is the act of first negotiating whether or not a concept is most appropriately represented in a text or non-text format and then developing a way to infuse that into the overall communication piece substantially more valuable for promoting effective conceptual development to a degree that warrants encouragement of *both* factors in instruction developing multimodal competency? Or, is the pairing of the two actions an unnecessary use of cognitive resources that does not dramatically affect student conceptual understanding?

Questions Related to Suggested Pedagogical Action

As mentioned earlier, regardless of the theoretical framework grounding the work, the authors in this book recognize that effective use of multiple representations or multimodal communication is not an automatic outcome of asking students to engage with multiple representations or to create multimodal communication. Some sort of instructional intervention, aimed at both encouraging the awareness of the utility of using different modes, along with recognizing the strategies that can allow for more effective multimodal communication, is needed. Again, to some extent, the pedagogical approaches can be categorized based on whether emphasis was on re-representations of similar concepts or creation of more broad based multimodal communication products.

In the cases where students were engaged in re-representational sequences of a targeted concept, the pedagogical sequences ideally facilitated a process of progressive refinement of a conceptual understanding modified to some extent by an awareness of the affordances and constraints of different representational tools. Students progressed through a sequence of presenting a current conception of a phenomenon using a particular modality. The sequential consideration of multiple representations allowed for re-inspection and re-evaluation of a current understanding. In being challenged to represent the idea in different ways, students were confronted with assessment of their own understanding as they attempted to reason through the strengths and shortcomings of different types of representations. As students refined their representations and searched for a more appropriate means of representing, more accurately, the targeted concept, the students were engaged in knowledge construction. The repeated cycle of evaluating a current representation in light of collected or provided data and observations and then considering whether or not a more appropriate means of representing the concept should be enacted helped students clarify their scientific understanding. Reflection after the fact allowed for a consideration of the overall changes in understanding. The pedagogy associated with this process involved carefully constructing opportunities both for students to encounter data and observations and then evaluate the meaning of the observations relative to the representations they had previously created. Attention was directed at the appropriateness or lack of appropriateness of different, progressive representations to help students understand that as our knowledge of phenomenon is improved or added to through scientific processes, our ability to more accurately represent natural phenomenon is improved as well.

In the cases in which students infused modes outside of text with text to describe related concepts, the pedagogical approaches discussed differed. In these cases, the goal of creating effectively integrated and linked communication pieces was emphasized in instruction aimed at helping students realize the effectiveness of first using modes other than text, and second, purposefully utilizing strategies to link those modes to the text and the other modes used throughout the entire piece of communication. In these situations, students were still considering their understanding of a concept, but they were being asked to communicate about that understanding by simultaneously linking different modes that described or detailed different aspects of the overall concept or theme. Instruction was less focused on the appropriateness of a specific mode but rather on the effectiveness of a unified, integrated communication piece in more accurately and appropriately describing the targeted information. The multiple representational aspect of this communication was not highlighted through targeted sequences of re-representation, but rather through the consideration of appropriate strategies to link different modes (often text and an alternative mode) representing the same concept together or to link different modes representing related but different ideas throughout a global communication piece together. The goal state of a well-integrated product was encouraged through instruction highlighting strategies available to link modes together, discourse aimed at justifying the use of particular modes and linking strategies, and the development of tools to assess the overall integration of multimodal communication.

The most fruitful question emerging from the consideration of the pedagogical approaches is how could aspects of all approaches be combined to better develop conceptual understanding, as a result of more effective use of modes as knowledge generation tools? Would a pedagogical sequence in which students are initially engaged in purposeful re-representation of a specific targeted concept help develop a sense of both the power of representations to progressively improve understanding as well as the strengths and weaknesses of different modes for communicating different characteristics of nature or for fulfilling specific purposes? Could this emphasis on re-representation be infused with purposeful discussion about strategy use to more explicitly and effectively link different modes together? Could this re-representational emphasis be progressively expanded until discussion and student engagement focused on summary type communication that included *both* multiple representations when appropriate and purposeful strategy use to create cohesive communication? Or, would the reverse of this sequence, in which students are first faced with the task of constructing more global communication pieces, and then, when in the course of the construction of these products they are asked to purposely link different modes together, they engage in discourse and activity surrounding the determination of what constitutes the most appropriate representation, be more productive? Research aimed at exploring the sequencing of a more "combined" approach could be beneficial in creating greater learning.

Questions Regarding Multimodal Communication and Student Reasoning

One of the hallmarks of the nature of science is that science knowledge claims are based on the accumulation and analysis of empirical evidence (McComas 2004). While many teaching approaches utilize visual modes of representation outside of text to present information to students, the "capacity to think with and through images" is typically less emphasized (Prain and Tytler 2012). In this way, teaching approaches often prioritize more formal verbal and linguistic logic over more informal, creative visual logic, even though scientists themselves often employ this type of reasoning (Prain and Tytler 2012; Gooding 2006). The use of multimodal communication and multiple representations described in this book are intended to engage students in opportunities to experience and employ both formal and informal reasoning facilitated through the use of different modalities.

Two main avenues of fostering reasoning can potentially arise from focusing on multimodal communication. First, the opportunity to create and critique representations enables students to reason through the process of attempting to match the most appropriate mode or combination of modes to best represent all aspects of a given concept. As students consider both their understanding of the concept and the affordances and constraints of the particular mode of representation they are creating or critiquing, the students generate new understanding of targeted concepts. As students compare personal representations to those of peers or those presented in accepted sources of scientific information, they potentially develop a representational competency (diSessa 2004) that includes both awareness of how the features of a particular mode are more adequate in representing a particular aspect or characteristics of a specific concept, and a more complete understanding of the concept itself.

A second reasoning opportunity arises from multimodal language engagement when considering the more global learning environment. Asking students to create and critique representations can be utilized as a method for synthesizing any of a number of classroom activities that have taken place. A representational task might offer students a chance to consider data collected and analyzed from hands-on experiments, information debated and discussed in classroom conversations, claims justified through presentation of evidence, ideas presented or enacted through demonstrations or kinesthetic role-playing, or information from any of a number of other classroom actions. Students must reason through the patterns that these varied classroom activities present, the degree of coherence between the different classroom activities, the differential level of evidence present for supporting different explanations, and the relationship between all of these and student prior understanding (Prain and Tytler 2012). Gooding (2006) points out that this not only provides an avenue for student reasoning, but again models the practices scientists utilize.

Through these reasoning processes, students can be made aware of the power of multiple representations and multimodal communication to not only describe but also to persuade. Students can participate in the scientific processes of utilizing communication tools and resources to both provide information about observations and data and to convince others that their analysis and interpretation of the data and observations is appropriate. In this way, the multiple representations and the multimodal communication can be considered as tools to aid in the critical scientific skill of negotiation. Facilitation of argument and debate, centered on the differential student interpretation of observations and data, and enhanced through the utilization of different forms of representation, is accomplished as a result of multimodal communication. One emerging question is to determine the best way to infuse the instruction about multimodal communication with typical classroom practices. More encompassing than the somewhat specific instructional pathways that are described in this volume to either encourage student use of strategies to link different modes together or to progressively move students through re-representation in different formats, the day-to-day pedagogical approach in the classroom sets the tone for way that the use of multimodal representations are viewed by students. If the progressions associated with multiple representations or multimodal communication are seen as independent learning progressions and not associated with the normal "way we do things in the classroom", students may not become aware of the potential to use this type of communication to reason and persuade. Students may see this as one of any number of specific instructional actions that they move on from when the unit of instruction has concluded. Conversely, if the use of multimodal communication and multiple representations is perceived as an infused and embedded aspect of the way activity in the classroom always proceeds, students may gain a greater grasp of the potential power of these resources and tools. For example, some preliminary pilot work has indicated that in classrooms where argument-based instruction is the norm, student gains from multimodal communication are enhanced. Further study is needed to attempt to clarify the link between the typical pedagogy and instructional practices of a classroom and the instructional and pedagogical practices that have been described in this volume as a means to promote student reasoning.

Measuring Multimodal Competency

The divergent approaches and studies reported in this volume are beneficial in terms of relating the wide range of positive pedagogical action associated with multimodal communication. Each study contributes to a full picture of the current state of research and serves to highlight the many different approaches educators can take to infuse multimodal work in their classrooms. However, in this chapter, we have viewed the studies holistically as an attempt to discover where the work converges to inform instruction. The ultimate point of convergence may be in attempting to determine what it means to possess and exhibit an appropriate multimodal competency. Different authors throughout this book have suggested different characteristics (i.e. "representational flexibility") that they deem critical for students to develop if the potential benefits of multimodal communication are to be obtained. While none of the individual studies may encapsulate the totality of a construct of true "multimodal competency", the studies taken together may suggest an avenue for defining this idea.

Obviously, the mode that is consistent throughout all of the studies presented is the use of text to describe scientific phenomenon under study. It would seem unlikely that having a multimodal competency, then, would not begin with an appropriate command of use of text to describe and communicate about scientific phenomenon. However, the use of text must be integrated with the use of appropriate modes that enhance the text as well as with appropriate strategies that link the text and the different modes together. Therefore, although true multimodal competency must include appropriate use of text, it likely also includes some, or all of the following abilities:

- Represent the same concept in multiple ways and recognize the strengths and weaknesses of the different representations.
- Represent multiple related concepts with an array of different modes, including text, and effectively link the modes together so that the reader is aware of the connections between the different modes.
- Recognize which concepts are not easily, or accurately described in text and therefore require representation in a mode outside of text.
- Apply strategies that explicitly link the text used to the modes outside of text used and the ability to tailor these strategies to both situations when text and other modes are communicating about different concepts and when they are used to reiterate information about the same concept.
- Create multiple, sequential representations of the same concept that progressively develop and clarify understanding of the concept.
- Explain why a particular type of mode was a better representation of a particular concept or idea.
- Recognize when a cluster of several different modes is necessary to appropriately and accurately convey information about a particular concept.

The aforementioned characteristics are all related to the actual production of the multimodal communication. In addition to these, a high level multimodal competency may actually include characteristics related to the epistemological outlook that the student holds. These characteristics may include the ability to:

- recognize the utility of multiple modes of representation as a tool to reason through the different aspects of a complex concept.
- understand that representations are models of our understanding of nature and as such, may be approximations that have limits, just as our understanding of nature has limits.
- understand that just as the knowledge we generate about science is the product of analysis and interpretation of data gathered from systematic testing of the natural world, the representations that we generate to communicate about nature are the products of this same process.

- recognize the utility of different modes of representation as tools to further our understanding of nature.

Two natural questions arise from the preceding lists. First, the obvious question of which of these characteristics is a critical feature in a multimodal competency. Secondly, however, is the question of how to measure this competency. While some of the studies here have described approaches to evaluating the level of multimodal competency a student has attained as an aspect of determining how beneficial the process is to the student, no studies have carefully conceptualized a thorough description of what a complete multimodal competency includes. This work will likely involve not only the evaluation of student work, but also work of experts in the different fields of science, as well as qualitative descriptions of the different multimodal products created and the rationale and reasons for creating the products as they did.

The studies presented in this volume provide evidence of benefit of student engagement in multimodal communication in varied international settings, with varied instructional practices, focusing on varied targeted concepts, and implemented by various teachers. The studies also highlight the wide range of specific curricular goals that can be accomplished when the use of multimodal communication is promoted as an epistemological tool and a tool for learning, rather than as solely a tool for communication. However, the most effective, efficient, reproducible, and generalizable pathway for accomplishing all of the curricular goals suggested throughout the book is far from clear. The questions raised in this chapter will hopefully drive new and dynamic research projects that further clarify the work that has begun as researchers continue the search for the most effective means for developing true scientific literacy for students. These projects must combine information from researchers, teachers, and students themselves as these groups seek to clarify the most effective pedagogical progressions, infused with the most appropriate student tasks, to encourage the most beneficial cognitive activities.

References

Anderberg, E., Svensson, L., Alvegard, C., & Johansson, T. (2008). The epistemological role of language use in learning: A phenomenographic intentional-expressive approach. *Educational Research Review, 3*, 14–29.

diSessa, A. (2004). Metarepresentation: Native competence and targets for instruction. *Cognition and Instruction, 22*(3), 293–331.

Gooding, D. (2006). From phenomenology to field theory: Faraday's visual reasoning. *Perspectives on Science, 14*(1), 40–65.

Gunel, M., Hand, B., & Gunduz, S. (2006). Comparing student understanding of quantum physics when embedding multimodal representations into two different writing formats: Presentation format vs. summary report format. *Science Education, 90*(6), 1092–1112.

Hubber, P., Tytler, R., & Haslam, F. (2010). Teaching and learning about force with a representational focus: Pedagogy and teacher change. *Research in Science Education, 40*(1), 5–28.

Klein, P. (2006). The challenges of scientific literacy: From the viewpoint of second-generation cognitive science. *International Journal of Science Education, 28*(2–3), 143–178.

MacWhinney, B. (2005). The emergence of grammar from perspective taking. In D. Pecher & R. Zwann (Eds.), *The grounding of cognition* (pp. 198–233). Cambridge, UK: Cambridge University Press.

Mayer, R. (1996). Learners as information processors: Legacies and limitations of educational psychology's second metaphor. *Educational Psychologist, 31*(3/4), 151–161.

McComas, W. (2004). Keys to teaching the nature of science: Focusing on NOS in the science classroom. *Science Teacher, 71*(9), 24–27.

McDermott, M. (2009). *The impact of embedding multiple modes of representation on student construction of chemistry knowledge.* PhD (Doctor of Philosophy) thesis. University of Iowa, http://ir.uiowa.edu/etd/253

McDermott, M., & Hand, B. (2013). The impact of embedding multiple modes of representation within writing tasks on high school students' chemistry understanding. *Instructional Science, 41*(1): 217–246.

Prain, V., & Tytler, R. (2012). Learning science in school through constructing representations. In C. Kennedy, & M. Rosengren, (Eds), *Spectra: images and data in art/science.* Proceedings from the symposium SPECTRA 2012, Canberra, Australia.

Sweller, J., van Merriënboer, J. J. G., & Paas, F. (1998). Cognitive architecture and instructional design. *Educational Psychology Review, 10*, 251–296.

Yore, L., & Treagust, D. (2006). Current realities and future possibilities: Language and science literacy – empowering research and informing instruction. *International Journal of Science Education, 28*(2), 291–314.

Yore, L., Bisanz, G., & Hand, B. (2003). Examining the literacy component of science literacy: 25 years of language and science research. *International Journal of Science Education, 25*, 689–725.